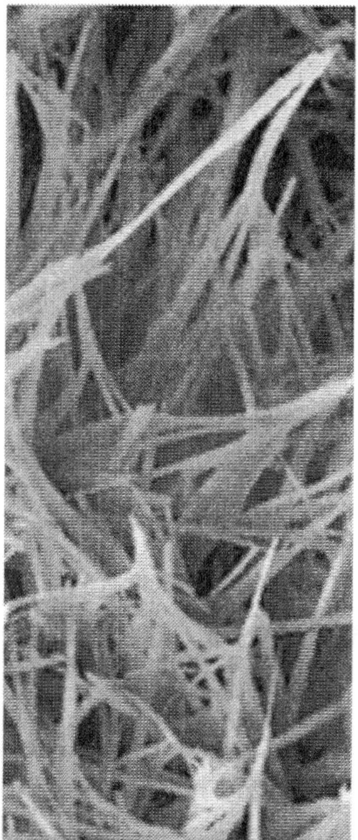

Materials Science of Concrete

SPECIAL VOLUME

Cement and Concrete—*Trends and Challenges*

Related titles published by The American Ceramic Society:

Materials Science of Concrete VI
Edited by Sidney Mindess and Jan Skalny
©2001, ISBN 1-57498-069-6

Materials Science of Concrete IV
Edited by Jan Skalny and Sidney Mindess
©1995, ISBN 0-944904-75-0

Materials Science of Concrete III
Edited by Jan Skalny
©1992, ISBN 0-944904-55-6

Materials Science of Concrete II
Edited by Jan Skalny and Sidney Mindess
©1991, ISBN 0-944904-37-8

Materials Science of Concrete I
Edited by Jan Skalny
©1989, ISBN 0-944904-01-7

Materials Science of Concrete —Special Volume: Ion and Mass Transport in Cement-Based Materials
Edited by R. Doug Hooton, Michael D.A. Thomas, Jacques Marchand, and James J. Beaudoin
©2001, ISBN 1-57498-113-7

Materials Science of Concrete —Special Volume: Role of Calcium Hydroxide in Concrete
Edited by Jan P. Skalny, Juri Gebauer, and Ivan Odler
©2001, ISBN 1-57498-128-5

Materials Science of Concrete —Special Volume: Sulfate Attack Mechanisms
Edited by Jacques Marchand and Jan Skalny
©1999, ISBN 1-57498-074-2

Materials Science of Concrete —Special Volume: The Sidney Diamond Symposium
Edited by Menashi Cohen, Sidney Mindess, and Jan Skalny
©1998, ISBN 1-57498-072-6

Boing-Boing the Bionic Cat and the Jewel Thief
By Larry L. Hench
©2001, ISBN 1-57498-129-3

The Magic of Ceramics
By David W. Richerson
©2000, ISBN 1-57498-050-5

Boing-Boing the Bionic Cat
By Larry L. Hench
©2000, ISBN 1-57498-109-9

Ceramic Innovations in the 20th Century
Edited by John B. Wachtman Jr.
©1999, ISBN 1-57498-093-9

For information on ordering titles published by The American Ceramic Society, or to request a publications catalog, please contact our Customer Service Department at 614-794-5890 (phone), 614-794-5892 (fax), <customersrvc@acers.org> (e-mail), or write to Customer Service Department, 735 Ceramic Place, Westerville, OH 43081, USA.

Visit our on-line book catalog at <www.ceramics.org>.

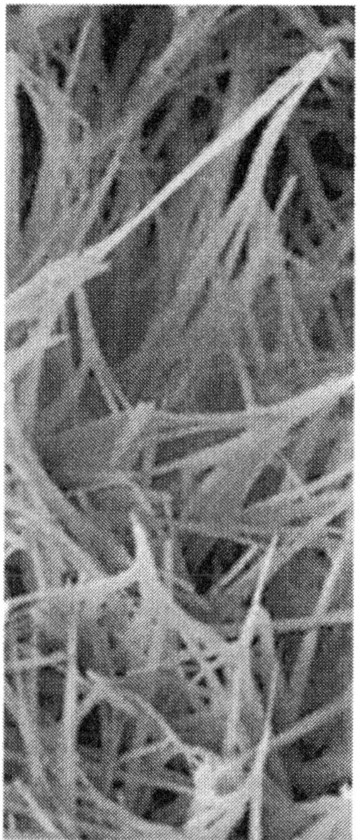

Materials Science of Concrete

SPECIAL VOLUME

Cement and Concrete—*Trends and Challenges*

Edited by
Andrew J. Boyd • Sidney Mindess

Series Editor: Jan Skalny

Published by
The American Ceramic Society
735 Ceramic Place, Westerville, Ohio 43081
www.ceramics.org

Proceedings of the Anna Maria 2001 Workshop on Cement and Concrete: Trends and Challenges. November 7–9, 2001, at Holmes Beach, Anna Maria Island, Florida.

Cover Photos, top row, l-r: "Storage of monosulfate in water at 22°C: formation of aggregates of parallel grown ettringite crystals" (fig. 19b); "Microstructure after 21 days hydration time at 5°C: hexagonal calcium-aluminate-hydrate" (fig. 17b); bottom row, l-r: "Microstructure after 2 h hydration time (low potassium content): Ettringite with the same morphology as after 10 min, increased syngenite formation" (fig 10a); "Formation of needle-like C-S-H phases during the hydration of C_3S. After 600 days hydration time the fibers grow up to a length of 1.5μm and the needle tips have a diameter of only approx. 5 nm" (fig. 4); "Microstructure after 2 h hydration time (high potassium content): Ettringite and syngenite with the same morphology as after 10 minutes" (fig. 10b) appear courtesy of B. Möser and J. Stark, from their paper "A New Model of Ordinary Portland Cement Hydration Derived by Means of ESEM-FEG," which begins on page 89.

Library of Congress Cataloging-in-Publication Data
A CIP record for this book is available from the Library of Congress.

For information on ordering titles published by The American Ceramic Society, or to request a publications catalog, please call 614-794-5890, or visit <www.ceramics.org>.

4 3 2 1–05 04 03 02

ISBN 1-57498-164-1
ISBN 13: 978-1-57498-164-3

Contents

PREFACE

This special volume of The American Ceramic Society's *Materials Science of Concrete* series represents a collection of presentations given and discussions held at the Anna Maria 2001 Workshop on Cement and Concrete: Trends and Challenges, held November 7–9, 2001, at Holmes Beach, Anna Maria Island, Florida.

The purpose of the workshop was to bring together representatives from different portions of the cement and concrete community (including industry, academia, and government) to discuss future trends and challenges in the field of concrete materials. Presentation and discussion topics included new developments and their potential for application, industry's view on what it needs from academia, academia's opinion on what it expects from industry and what it has to offer, technical advances and barriers to advancement, and the role of governments, business issues, ecological barriers, and societal barriers.

The workshop organizers and the editors of this special volume would like to acknowledge their appreciation of all the workshop participants for their active involvement in the deliberations, with special thanks to the invited speakers. The workshop would not have been possible without sponsorship by numerous governmental and industrial organizations, specifically the University of Florida, the National Institute of Standards and Technology, the National Research Council of Canada, and RILEM, as well as by California Portland Cement Company, Dyckerhoff Zement GmbH, Grace Construction Products, Holcim, Holderbank, Lafarge Corporation, Master Builders Inc., Materials Service Life, Portland Cement Association, and R.J. Lee Group Inc.

We would also like to acknowledge the generous support of the Hon. Carol Whitmore, mayor of Holmes Beach and her staff for allowing us to use the city hall conference room and for helping to create a pleasant workshop atmosphere. Thanks are also due to the reviewers of the submitted manuscripts for their thorough and timely reviews. Finally, in the name of all the authors and participants, we would like to compliment the professionalism of the staff of The American Ceramic Society. Their close cooperation with the editors enabled timely and quality publication of this special volume.

Andrew J. Boyd

Sidney Mindess

Jan Skalny

Sponsors and Donors
ORGANIZING COMMITTEE

Jan Skalny
Andrew J. Boyd
Sidney Mindess

SPONSORING ORGANIZATIONS

California Portland Cement Company, Glendora, CA, USA
Dyckerhoff Zement GmbH, Wiesbaden, Germany
Grace Construction Products, Cambridge, MA, USA
Holcim, Holderbank, Switzerland
Lafarge Corporation, Paris, France
Master Builders Inc., Cleveland, OH, USA
Materials Service Life, Pittsburgh, PA, USA
National Research Council of Canada, Ottawa, ON, Canada
National Institute of Standards and Technology, Rockville, MD, USA
Portland Cement Association, Skokie, IL, USA
RILEM, Paris, France
R.J. Lee Group Inc., Monroeville, PA, USA
University of Florida, Gainesville, FL, USA

WORKSHOP PARTICIPANTS

Carmen Andrade	Instituto Eduardo Torroja, Spain
Adolf Bajza	Slovak Technical University, Slovak Republic
James Beaudoin	National Research Council of Canada, Canada
Mike Bergin	Florida Department of Transportation, FL, USA
Andrew J. Boyd	University of Florida, FL, USA
Andrzej M. Brandt	Polish Academy of Sciences, Poland
David Cahn	California Portland Cement Co., CA, USA
Tate Coverdale	Master Builders Technologies, OH, USA
Norah Crammond	Building Research Establishment, UK
Sidney Diamond	Purdue University, IN, USA
Christos Drakos	University of Florida, FL, USA
Emery Farkas	Grace Construction Products (Ret.), FL, USA
Stephen Forster	Federal Highway Administration, VA, USA
Juraj Gebauer	Holcim Group Support, Switzerland
Doug Hooton	University of Toronto, Canada
Felek Jachimowicz	Grace Construction Products, MA, USA
Ole Mejlhede Jensen	Aalborg University, Denmark
James Kirkpatrick	University of Illinois, IL, USA
Lesley Ko	Holcim Group Support, Switzerland
Jacques Lukasik	Lafarge Corporation, France
Weiping Ma	Holcim, MI, USA
Jacques Marchand	Laval University, Canada
Sidney Mindess	University of British Columbia, Canada
Denis Mitchell	McGill University, Canada
Bernd Möser	Bauhaus-Universität Weimar, Germany
David Myers	Grace Construction Products, MA, USA
Jan Olek	Purdue University, IN, USA
Jim Pierce	Bureau of Reclamation, CO, USA
Ken Rear	Lehigh Cement, GA, USA
Jean-Claude Roumain	Holcim, CO, USA
Karen Scrivener	Swiss Federal Institute of Technology, Switzerland
Surendra Shah	Northwestern University, IL, USA
Jan Skalny	Materials Service Life, FL, USA
Åke Skarendahl	Swedish Cement & Concrete Research Institute, Sweden
Robert Suderman	Lafarge Canada, Canada
Joseph Tedesco	University of Florida, FL, USA
Paul Tennis	Portland Cement Association, IL, USA
Niels Thaulow	R.J. Lee Group, PA, USA
Pavel Trtik	University of Paisley, UK
Danica Turk	U.S. Gypsum (Ret.), FL, USA
Hélène Zanni	École Supérieure de Physique et de Chimie Industrielles, France

21ST CENTURY CEMENT AND CONCRETE RESEARCH: WHERE DO WE GO FROM HERE?

Sidney Mindess
Department of Civil Engineering
University of British Columbia
2324 Main Mall
Vancouver, BC, V6T 1Z4, Canada

ABSTRACT

Over the last century, we have made great progress in cement and concrete research and development. We can now produce high strength, durable, and self-compacting concretes, made possible by our better understanding of the hydration process, and by advances in admixture technology. From reading the recent cement and concrete literature, it would appear that all of the significant problems have already been solved. The questions then remain: Why do we still continue to produce concrete that is not durable, that cracks excessively, and that is environmentally unfriendly? Why do we still have concrete failures? This presentation will focus on some of the reasons for this sorry state of affairs (e.g. inadequately trained engineers, poorly trained workforce, insufficient transfer of technology from the researchers to the field, politicization of professional organization, the litigation industry, and so on). It will conclude with some suggestions for both research and practice that may help to alleviate some of these problems.

INTRODUCTION

The 20th Century saw tremendous progress in concrete research and development. In 1901, cement and concrete were relatively primitive materials. Despite the seminal work of Le Chatelier, there was little general understanding of the chemistry of Portland cement, and even less of the structure of the hydrated cement paste.* Portland cement was thought to consist primarily of C_3S, C_3A, and "a fusible calcium silico-aluminate"; others believed that the alumina occurred mostly as C_2A. The precise role of each of these constituents in the behavior of the cement was still not known. As well, the "chemical changes that take place in the setting and hardening of cements are not thoroughly understood." The chief hydration products were considered to be "hydrated monocalcic silicate and calcium hydrate" ($CaO \cdot SiO_2 \cdot H_2O$ (?) and $Ca(OH)_2$), plus some ill-defined hydrated calcium aluminates. The common view was that "these chemical changes cause the formation of crystals which by their interlocking and adhesion give strength to the new compounds."

The cost of these Portland cements was in the range of $2.00 to $3.00 per barrel* (about $10.60 to $15.90 per ton).

There were as yet no standard specifications for Portland cement; it was defined by the Association of German Portland Cement Manufacturers as "an article as is made by calcining a thorough mixture, consisting essentially of calcareous and clayey substances, and then grinding the same to the fineness of flour." Chemical tests of portland cement were thought to be of not much value (a view which some modern engineers still share), though Sabin felt that "it is not impossible, then, that chemical tests may yet play a more important rôle in cement testing, especially if the method of analysis can be made more simple and rapid."

To characterize the mechanical properties of hydrated cement, the tensile test (!) was the one most frequently used. It was considered that "although tests of compressive strength are of interest from a scientific point of view, it is not considered that they would give much greater information ... than is given by tensile tests, and therefore they need not be included in an ordinary series of acceptance tests." Indeed, it was a test for *adhesion* (bond strength between mortar and stone) that many considered to be the most rational "rupture" test.

There were, as well, other tests available for determinations of the fineness of cement (by sieve analysis) and cement soundness, but none of these was then standardized.

Turning now to concrete, the significance of the water/cement ratio was still not understood**, and so mix proportioning was an even more arbitrary procedure than it is now. There were no commonly used mix design criteria, in terms of either strength or durability. No workability tests were specified, and so workability could only be described qualitatively (e.g. "the mortar was wet enough to quake like liver under moderate ramming"). Compressive strength tests (for comparing different cements or concrete, but not for specifying concrete for a specific project) were carried out on cubes of various sizes, both 6-inch and 12-inch cubes being common. Needless to say, neither physical nor chemical admixtures were available.

The cost of this concrete (in place!) was typically $4.50 to $5.50 per cubic yard. and, while "concrete-steel" (i.e. reinforced concrete) combinations had been used for special purposes for about 25 years, they were only now coming into common use. Simplified methods for analysis and design were available for reinforced concrete beams, but not for other types of structural members.

MODERN CEMENT AND CONCRETE SCIENCE AND TECHNOLOGY

Over the past 100 years, there have of course been tremendous advances in cement and concrete technology. We now have essentially a complete understanding of the chemistry of Portland cement itself, and the roles of its constituents. We can "tailor-make" cements with various properties, and there are no end of test specifications for both the composition and physical and mechanical properties of cements. The mechanical properties are characterized by compressive tests on mortars, while the composition is, as Sabin predicted a hundred years earlier, controlled by chemical tests and specifications.

We also have a pretty detailed understanding of the hydration process and the structure and properties of the hydration products, even if there are a few unsolved questions. Similarly, we have an excellent understanding of how to make good concrete, with the desired mechanical and durability properties, based largely on applications of the w/c ratio law, of Feret and Abrams. We can optimize concrete mixes for almost any materials and/or environmental exposures. As well, there are a whole host of standard test methods for the properties of concrete. We also now have detailed methods for the design of reinforced concrete ("concrete-steel") structures. Using finite element methods and/or fracture mechanics, we can even predict and analyze cracking and crack growth in concrete.

Finally, we now have available a large number of admixtures (air-entraining admixtures, superplasticizers, corrosion inhibitors, supplementary cementing materials, etc.) and other additions (e.g. fibers) with which we can modify the properties of concrete almost at will. For instance, we can now produce:

- concretes with compressive strengths of up to 600 MPa; concrete strengths greater than 100 MPa are becoming commonplace
- self-compacting concretes
- corrosion inhibiting concretes
- "tough" concretes (with the use of fibers)
- durable concretes
- economical concretes

Indeed, from a reading of the current journal literature, and the literally thousands of pages of conference proceedings published each year, one would be forced to conclude that we have "solved" all of the problems that matter regarding cement and concrete; all that remains is a few loose ends to tie up.

THE CURRENT STATE OF PRACTICE

Given the above, one would think that concrete problems in engineering practice would be uncommon, and that the problems that did occur would be primarily due to human error. The questions then arise: Why do we still so frequently have durability problems, severely cracked or spalled concrete, an environmentally unfriendly material, and concrete failures? Why is there such a wide gulf between what we *know* how to do properly, and what we *actually do* in practice? Why is construction litigation such a growth industry?

i) Durability

Today's concretes are in general less durable than the "old" (pre-1960) concretes, for several reasons:

- higher C_3S contents
- more finely ground cements
- too much reliance on water/cement ratio considerations, and too little consideration of the total water content.

ii) Cracking

We can produce virtually crack-free concrete by the proper use of a combination of fibers, shrinkage reducing admixtures and proper curing techniques. Nonetheless, we commonly see

- plastic shrinkage cracking due to improper (or no) moist curing
- cracks due to improper placement and/or compaction
- cracks due to the more "brittle" nature of modern concretes

iii) Environmental considerations

The production of Portland cement generates a large amount of CO_2 emissions (1 ton cement $\cong$ 1 ton CO_2). However, at least in North America, there is still insufficient use of supplementary cementing materials, though there are some promising developments in this area. As well, there is insufficient recycling of demolished concrete.

iv) Concrete Failures

A few hours of browsing through the web will turn up thousands of examples of failures of concrete or concrete structures. These are due, in whole or in part, to some combination of material deficiencies, inadequate design, or poor construction practice.

WHERE DO WE GO FROM HERE?

There are, of course, no simple answers to the questions posed above. But it is clear that, despite our highly sophisticated research, we still lack a number of the requisites for the consistent production of high quality concrete:

- A properly trained cadre of engineers specializing in cement and concrete technology. In North America, for instance, we simply do not train such people at the undergraduate level in our universities.
- A properly trained work force.
- A much better transfer of both knowledge and technology from the laboratory to the field. Our usual approach to this – publishing research papers in the standard journals – appears not to be very effective in this regard.

There is, unfortunately, no "quick fix" for this state of affairs. However, there are a number of things that we can do to begin to turn things around. Amongst other things, we need:

1. a better molecular-level understanding of the chemistry of cement hydration, and of chemical attack mechanisms, so that we can better deal with durability problems, and with admixture development;
2. concretes with better dimensional stability;
3. tougher, less brittle materials;
4. smart "self-policing" materials, which will somehow "tell" us when codes or specifications are being violated;
5. to establish one (or more) university undergraduate programs specifically directed at producing cement and concrete engineers. This would be quite different from the standard Civil Engineering programs from which most engineers dealing with construction materials are drawn.
6. an improved public perception of the cement and concrete industry, and a very much higher profile for cement and concrete research and development;
7. better ways of disseminating information to structural engineers, geotechnical engineers, and others involved in specifying or using concrete.

As well, we somehow have to reorient our approach to cement and concrete research, so that the academic research focuses on fundamentals, and "applied" research is mostly left to industry. Unfortunately, the following verse (Ecclesiastes I:II) is all too applicable to much current concrete research:

"There is no remembrance of former things; neither shall there be any remembrance of things that are to come with those that shall come after."

That is, too much of our current research is repetitive of things that we have long known, and as such is eminently forgettable.

Finally, we have to re-examine the way in which we write codes and specifications. Historically, the traditional "consensus standards" approach has served us well. However, it appears to be becoming increasingly "politicized," in part due to the increasing commercial pressure to write standards that appear to favor one material or brand over another, and in part due to the increasing amount of construction defect litigation.

CONCLUDING REMARKS

Despite the earlier tone of this diatribe, there are, in fact, many fine researchers in both industry and academia. There is innovative and useful current research on cement, concrete and concrete construction techniques. However, we hold a view of research that seems to have descended from the "invisible hand" theory of Adam Smith; namely, we assume that if we all follow our individual research paths, we will somehow (and inevitably) carry out the research that meets our most pressing industrial and national strategic objectives as well. Unfortunately, this is not usually the case. It is my hope that the discussion above might cause at least a few of us to rethink our research priorities, and the way in which we try to transmit our findings to the profession at large.

Photo 1: S. Mindess

Photo 2: E. Farkas and A. Bajza

Photo 3: J. Olek, A.J. Boyd and S. Mindess

Photo 4: N. Crammond and N. Thaulow

FROM MATERIAL TO THE STRUCTURE: FROM THEORY TO APPLICATION

Carmen Andrade
Institute of Construction Science "Eduardo Torroja" of the CSIC of Spain.
Serrano Galvache, s/n
28033 Madrid, Spain

ABSTRACT

At this dawn of a new century, reflection on the future trends that can be imagined in the development of materials and structures is very pertinent.

In this paper, the aspects addressed are the following: a) comments on the management of R&D activities related to the process of innovation, b) the benefits of leadership, c) the organization of research units and d) the role of international associations, in particular Rilem.

INTRODUCTION

Exploring future trends by analysing present data should be a permanent exercise for researchers. To make this exercise at the dawn of a century is particularly pertinent.

Underlying research activities exists an organization that assures the efficiency of the invested efforts, and facilitates the transfer of knowledge to industry and to practice.

In this paper, several aspects related to the organizational tools for optimising this transfer are discussed. First, some considerations on the basic management of R&D activities are made, followed by a stressing of the benefits of leadership. Afterward, the importance of having well-structured research units is emphasized, and finally comments on the important role of international associations, taking Rilem as an example.

MANAGEMENT OF R&D ACTIVITIES

Due to the known fragmentation of the construction industry, the main challenge related to R&D activities is the integration of the whole process of innovation. That is, to integrate material level research with that at the structure level and also integrate research with application. The management of these activities is not an obvious matter in the case of the construction sector.

Figure 1 shows the process of innovation as a sequential or "chain-like" organization. Thus, starting with the research (either basic or oriented) the consecutive steps may be summarized as: development, innovation and marketing. In each of these steps a different profile of worker or department is necessary. This chain is the more generally used model and it seems logical. However, its main disadvantage is that a break or inefficiency in one of the chain-steps leads to a failure of the whole process. This is what generally happens at present when academic or oriented research is transferred to an industrial department, which in turn has to contact the production and marketing areas in order to make the innovation a real product.

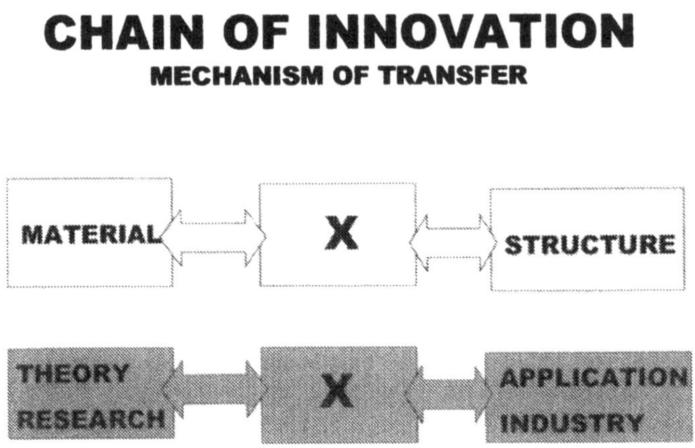

Figure 1

Figure 2 shows the circle of innovation that is a more efficient research management model, though more difficult to create and maintain. In this model, the research is the core or nucleus of an outer circle that encompasses the multiple participants or actors of the innovation process: industrial departments, professionals, governmental authorities, etc. The transfer (represented by arrows in the figure) is made between two partners through the shortest route, usually with the research unit being the catalyst. This kind of management better assures that a broken contact does not lead to the failure of the whole process. In practice, this means that research units or centers should be organized to have Boards in which the different partners of the innovation process (and not only industry) should be present. It also calls for multidisciplinary units having material scientists and structural engineers working together in a multi-material approach.

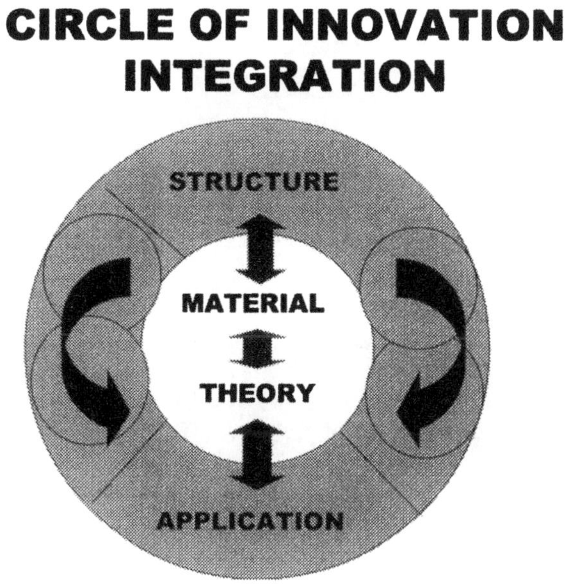

Figure 2

BENEFITS OF LEADERSHIP

The past century has produced very prominent personalities among civil engineers and architects that have made outstanding contributions to the development of cement and concrete technologies. Thus, Freyssenet, Leonhardt, Nervi, Wright and Torroja have exerted crucial leadership. Since the author of the present paper knows Eduardo Torroja's life better than any other, it will be taken as an example to illustrate the need for leadership of this kind in order to maintain the respect of society for concrete structures.

Eduardo Torroja was born in 1899 and passed away in 1961. Very recently, several events have been organized, primarily in Spain but also in many other countries, to commemorate the 100th anniversary of his birth. During his relatively short life, Torroja played all the roles that an engineer can. Originally a designer and constructor, he noticed the need to base design on well-founded principles, which led to very durable structures (Figure 3).

1933

ALGECIRAS MARKET
48 m span – 9 cm thick
Eduardo Torroja

Figure 3

In 1934, he founded, along with other colleague architects, the Technical Institute of Construction, which is the antecedent of the present Institute of Construction Science "Eduardo Torroja" IETcc. The Institute was private until 1940, when it was incorporated into the newly created National Research Council, a public entity. In 1949, the Institute of Cement was merged with that of Construction, and Torroja became Director until his death (Figure 4).

During this period in his life, Torroja continued to be a designer, though with less intensity, and became a researcher and professor at the university, integrating all of the activities and partners relevant to the innovation process into a research center. He was cofounder of CEB, Rilem, CIB, FIA, IASS and UEAtc. Examples of his multi-material and multidisciplinary approach are his publications. The most famous of these is "Philosophy of the Structures" [1] where he deals with structural design without needing any mathematical equations. This book has been translated into numerous languages, the most recent being Korean. Examples of other publications are given in References 2-7. It can be appreciated that he was working not only with concrete but also with steel and masonry. Also notable was his contribution, as chairman of the CEB committee on safety, to the definition of partial safety factors, which allowed designers to treat safety of structures in a simple enough manner to be included in standards.

Figure 4

The time needed for an idea to reach practice was very short, since the IETcc worked as the core of a multidisciplinary activity that had representation on its Board from the main industrial sub-sectors, as well as the main governmental Departments.

The influence of Torroja on Spanish and South American engineers and architects was not only extraordinary at his time, but still persists. As a result, concrete is the structural material of choice not only for public works but also for houses in Spain and Iberoamerican countries

The structures designed and built by Torroja still perform with remarkable durability. Shown in Figure 5 is the "Zarzuela" Hypodrome in Madrid, built in 1935 and having several reinforced concrete shells 7cm thick in cantilever. Figure 3 shows the dome of the "Algeciras" Market in the south of Spain, built in 1933 only 300m from the sea and having some chlorides mixed in due to the use of beach sand. It has a span of about 48m of and is 9cm thick. After about 65 years of life, both structures show only minor deterioration in spite of the fact that the concretes had between 15 and 18% porosity. The majority of the reinforcement in both structures exhibits no corrosion.

Hipodrome of Zarzuela 1935 - Madrid

Figure 5.

RESEARCH UNITS

Regarding the Institute that E. Torroja founded, and which took his name after his death, it still maintains the original multidisciplinary approach (structural engineering, material science, architecture and building physics). It pursues activities ranging from oriented research to technological development, assessment reports and standardization (Figure 6). It has a Board where the major industrial organizations and related governmental departments are represented.

The IETcc continues to play a leading role in research related to construction in Spain and Iberoamerica, due mainly to its flexible and versatile organization following the innovative circle model, which has demonstrated its efficiency in spite of more than 60 years of existence. It facilitates: a) the quick interaction between material, structure and building and b) the quicker transfer from theory to application.

INTERNATIONAL ASSOCIATIONS

Another important tool for facilitating this transfer is the International Associations. They are, in general, a good platform for multidisciplinary exchange. International Associations actively operating in concrete structures are:

Rilem, FIB, IABSE, CIB, IASS and many other more regional or specialized entities.

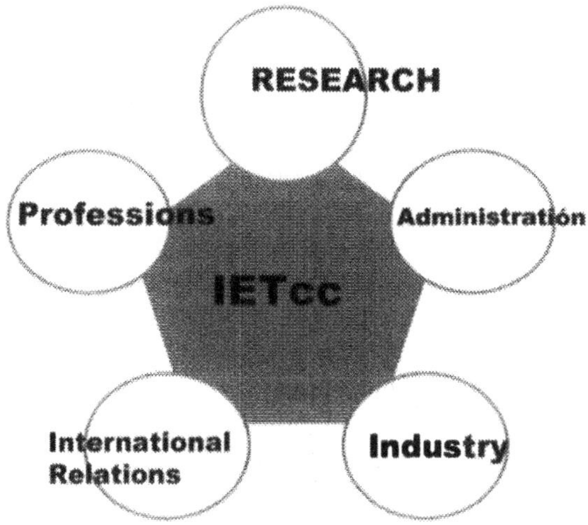

Figure 6

Rilem is the most familiar to the author of the present paper and therefore will be taken as an example. Rilem is presently defined as the International Union of Laboratories and Experts in building materials, systems and structures. Its mission is to advance scientific knowledge in construction materials, systems and structures and encourage the transfer of such knowledge worldwide by promoting networking.

Rilem was created in 1947, after World War II, to harmonize testing methods. Due the intensive reconstruction of a Europe destroyed by war, new materials and construction procedures were appearing without enough guarantee of good performance. Rilem played an important role in unifying testing methods, among which are: the slump test,[8] the compression test,[9] the splitting (Brazilian) test [10] and many other that many do not remember. All were developed under Rilem's framework.

Due to the progressive activity of ISO at the international level and of ASTM at the North American level, Rilem's role appears to be less necessary. On the other hand, the need to understand the fundamentals for producing reliable testing methods has changed the profile of Rilem members from directors of the centers to more pure researchers, mainly material science specialists.

At present, Rilem primarily gathers academics and researchers of public and private centers that meet not only to put together existing knowledge, but also to create new knowledge through their collective work.

Rilem works by placing the authority on the General Council (GC), which is composed of members of Standing Committees, national delegates, and representatives from the prior year's Technical Committee chairmen. Figure 7 shows the present organization. The GC has an executive body, the Bureau, in addition to which three Standing Committees operate: the TAC (Technical Activities Committee), the MAC (Management committee) and the DB (Board of Directors), all assisted by the General Secretariat.

The TAC coordinates and stimulates the technical and scientific work through the recent creation of four "clusters", which group technical committees of similar thematic content and are chaired by a Convenor. The four present clusters are:

A. Materials Characterisation/Microstructure
B. Properties Evaluation and Processing
C. Design and Service Life
D. Performance and Deterioration Mechanisms

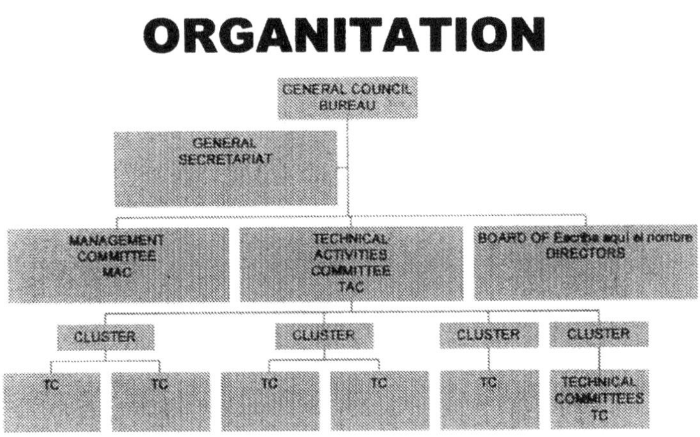

Figure 7

The products of the TC's (which cannot stand longer than around 5 years with the same subject and membership) are:

- State-of-the-Art reports
- Proceedings of workshops and symposia
- Testing methods
- Rilem recommendations

Other Rilem publications are the Journals, which at present are: *Materials and Structures* and *Concrete Science and Engineering*.

Recently created, the Board of Directors (DB) will take over part of the role that the Directors of Research units had in the past. Management of research in this sector is complex, and exchange of experience will contribute to a higher efficiency of the so-called "Titular members" of Rilem.

The first task undertaken by the DB has been the preparation of a "white paper" on Research priorities in the Construction sector, which could serve to stimulate the interest of society for research activities in the area. It is very much in line with the aims of the present workshop.

This white paper will first be elaborated internally within the DB, during 2002, and then a workshop and an international seminar will be organized to elaborate on the final version in collaboration with all other international, regional or national organizations, authorities and professionals interested in R&D. At the European level, the General Directorate of R&D has welcomed this initiative of Rilem and may sponsor the planned workshop and seminar.

In September 2001, the Rilem delegates in Japan (Dr. Motohashi) and in the USA (Dr. Shah), together with Mr. Pero of the GD for R&D of the European Union, presented papers to the General Council concerning research priorities. The intention is to invite speakers from other geographic regions in order to collaborate on the elaboration of the mentioned "white paper" in September 2002.

REFERENCES

1. E. Torroja, "Philosophy of Structures (Univ. Berkeley, 1958).
2. E. Torroja, "L'emploi des cables d'acier dans les constructions en béton armé (Vol. del Congrés des Ponts et Charpentes, Viena, 1929).
3. E. Torroja, Le voile mince du Fronton Recoletos à Madrid (Association Internationale des Ponts et Charpentes, Zürich, 1938).
4. E. Torroja, "Planning and Execution of Structures of Reinforced concrete in conjunction with Arc Welded Steel", (Premiada en concurso Internacional, Gotemburgo, 1949.

5. E. Torroja, "General report on safety" (IV Congress I.A.B.S., Cambridge, 1952)
6. E. Torroja, "Wirtschaftliche Gestaltung von Staumauern durch Modellversuche" (Beton und Stahlbau, I, 1955).
7. E. Torroja, "Deformation Processes and Rheological Patterns" (Symposium su la plasticitá, Varenna, 1956).
8. Rilem – Technical Recommendations for the Testing and use of Construction materials – CPC2.1 – 1975 – Slump test of concrete – E&FN Spon Editors pg. 4.
9. Rilem – Technical Recommendations for the Testing and use of Construction materials – CPC4 – 1975 – Compression Test of Concrete – E&FN Spon Editors pg. 4.
10. Rilem – Technical Recommendations for the Testing and use of Construction materials – CPC6 – 1975 – Tension by splitting of concrete specimens" – E&FN Spon Editors pg. 4.

Photo 5: C. Andrade

Photo 6: D. Turk

Photo 7: T. Coverdale and F. Jachimowicz

Photo 8: L. Ko, J.C. Roumain, J. Lukasik and J. Gebauer

NUCLEATION AND GROWTH IN CONCRETE RESEARCH

Ole Mejlhede Jensen
Per Freiesleben Hansen
Build. Techn. & Struct. Engin
Aalborg University
Sohngaardsholmsvej 57
9000 Aalborg, Denmark

Gunnar Idorn
International Consultant
Tovesvej 14B
2850 Nærum
Denmark

ABSTRACT

Concrete has a short history. Nevertheless, today it is the most widely used construction material, and indispensable for global welfare development in our century. The basic elements to this technological success have been low price and the nearly unlimited compositional, geometric and structural freedom that concrete allows. The development of concrete technology is characterized by a series of innovative steps during the last 150 years. These steps concern material development, for example reinforcement and super plasticizers, but also development of the process technology such as design and monitoring of concrete curing processes. The progress of concrete technology has been driven by scientists' curiosity and industrial needs, and has been enabled through the development of still more sophisticated scientific instruments for observation and analysis of concrete; concrete technology and concrete science are interdependent and inseparable. Innovative steps will surely continue to develop concrete technology in the future. In addition to scientific curiosity and industrial needs, future development will most likely also be severely influenced by environmental demands – a call for "green concrete". This, however, is a positive issue for concrete research since, essentially, concrete is born out of nature.

CONCRETE – A GENIUS MATERIAL

Traditional concrete seems to be a poor material when its properties are superficially compared with other materials like wood, steel and brick. From a construction point of view, traditional concrete has, for example, low strength, low fracture toughness, and a low thermal shock resistance[1], see Figure 1.

A more detailed look at concrete, however, reveals that it is a genius material. For example, concrete is highly shapeable. This enables a considerable increase in material performance. In many design situations, the properties of concrete can be utilized much more efficiently than the properties of wood, for example. The dotted displacement in Figure 1 shows how concrete strength, in reality, can be regarded when shapeability is taken into consideration. In addition, the shapeability of concrete allows a structural "freedom of design" which is unique compared to other materials.

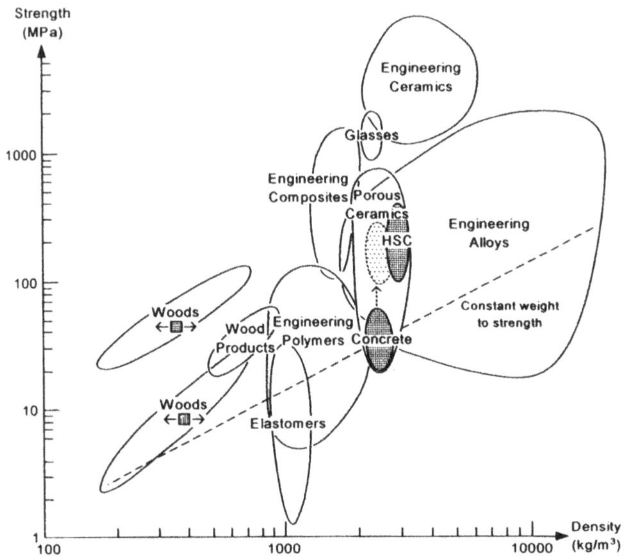

Figure 1. Traditional concrete has relatively low strength from a density point of view: All materials above the broken line are relatively stronger than traditional concrete. Modern, high-strength concrete, HSC, however, is very different from traditional concrete. For significance of the dotted displacement, see text. Based on Ashby[1].

Relative cost is another highly favorable property of traditional concrete. As shown in Figure 2, most other materials have a higher price to strength ratio. The low price of concrete is based on the abundance and easy retrieval of raw materials for concrete, e.g. limestone, aggregate, gypsum and fuel.

The low price and the geometric and structural freedom of design that concrete allows are important elements in its success. However, an additional peculiarity has been perhaps even more important: the composition of concrete can be varied to a significant degree, enabling an extreme variation of concrete properties to be developed and controlled. A notable example of this was the development of high-strength concrete, HSC, which may possess strengths 10 times higher than traditional concretes (see Figure 1). HSC can be considered an *artificial rock* that is as strong as many natural rocks[2]. In this respect, HSC can also be considered an even further improvement of the *artificial stone* for which Aspdin was granted a patent[3] in 1824.

The material price of ultra HSC may exceed US\$2000 per m^3 – up to 50 times the price of traditional concrete. The price of HSC, however, is not a limiting factor for a number of applications since HSC is still cheap compared with other materials (see Figure 2).

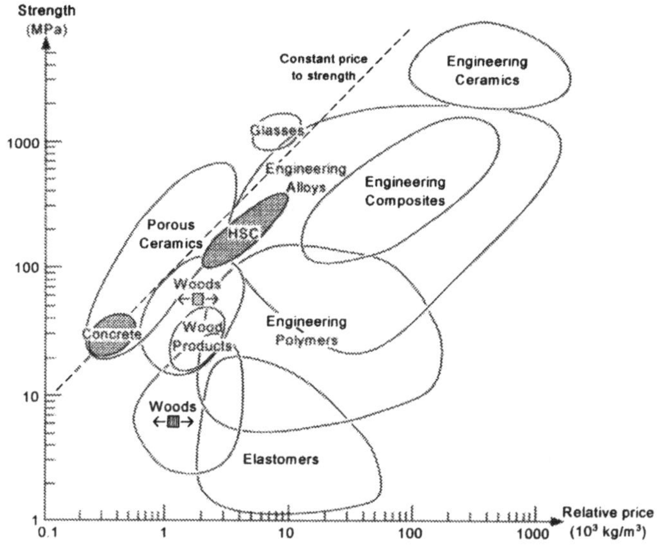

Figure 2. Concrete is an economic material with regard to strength. Based on Ashby[1].

The development of HSC is just one example of the very significant progress concrete technology has undergone since Aspdin's invention of Portland cement in 1824. A major part of this progress is attributed to a series of innovative milestones as will be exemplified in the next section. This progress has occurred within all parts of concrete technology: Material technology, process technology and construction technology. Examples of *material technology* progress include development of synthetic, lightweight aggregate, polymer impregnation, fiber reinforcement and mineral and chemical admixtures, which have revolutionized the possibilities of concrete technology. Pumping, vibration, ready-mix and development of curing technologies[4] are examples of progress *within process technology*, whereas prestressed concrete, shell structures and building systems with precast concrete components are improvements of the *construction technology*.

Forces behind these changes have been scientific curiosity and market needs for better performance and better economy. Today, global issues such as environmental problems have also been put on the agenda by international society.

THE HISTORY OF CONCRETE REINFORCEMENT – SO FAR

A significant drawback of unreinforced concrete is its low tensile strength. The tensile strength of concrete is only about 5-10% of its compressive strength. In addition, concrete is – like other ceramics – brittle. This leads to a wide scatter in tensile strength and severely limits the applicability of plain concrete.

The idea of reinforcing brittle materials is very old. To increase the free span of marble stones in Acropolis (approx. 440 B.C.) iron rods were cemented within grooves in the marble[5]. However, it took a long time before this idea was applied to concrete. The first person to use iron seriously as reinforcement of concrete was not an engineer; it was the French gardener, Monier, who made large tubs for orange trees by embedding a mesh of thin iron rods in concrete in 1849. In 1867 Monier received a patent on this idea and later Monier also used reinforced concrete for pipes, railway ties, floors, stairways, bridges and buildings.

Monier, however, was not successful in his promotion of reinforced concrete. It was not until about 1890 that major use of reinforced concrete took place, and that was primarily due to the further technical development and promotion by two engineers, the Frenchman Hennebique and the German Wayss. The price reduction and improvement of iron and steel during the nineteenth century also promoted this development. In Britain, during the reign of Queen Victoria (1837-1901), the price of steel was reduced more than tenfold[5]. Around 1900, reinforced concrete became a serious competitor to wood, steel and masonry for structural applications in Europe[6]. In the USA, this development was delayed due to a powerful steel industry and conservative civil engineering education. Today, 150 years after its introduction, reinforcement of concrete is considered a matter of course; reinforced concrete is merely referred to as concrete (see Figure 3).

However, the idea of reinforcing concrete has been developed even further. In normal reinforced concrete, the concrete in the tensile zone does not contribute to strength but it does add to the dead load. This is a disadvantage, especially for longer spans. The invention of prestressed concrete enabled much more efficient structures to be built. In prestressed concrete, high tensile steel cables are stretched and fixed in the concrete. This creates compressive stresses throughout the concrete, and leads to a more efficient use of the materials. Prestressed concrete was introduced about 1890, but this concept was very slow in catching on. Successful further development of the technique was done by the French engineer Freyssinet around 1930-1940, and this was the basis for the worldwide application of prestressed concrete around 1960. Today, Freyssinet is considered the father of prestressed concrete.

Another important development in reinforced concrete is fiber reinforcement. Efforts to incorporate fibers into concrete have been attempted since 1900. However, it was not until research was carried out around 1960-1970 that significant development of fiber reinforced concrete took place. This was based on fracture mechanics concepts and modern composite theory. A further increase in the use of fiber reinforcement occurred after 1990. This was due to the introduction of new fiber types and fiber geometries in the late 1980s (see Figure 4). Fiber reinforcement has increased the use of concrete. Applications of fiber reinforced concrete include pavements, thin shells, tunnel casings and many different repairs[7,8].

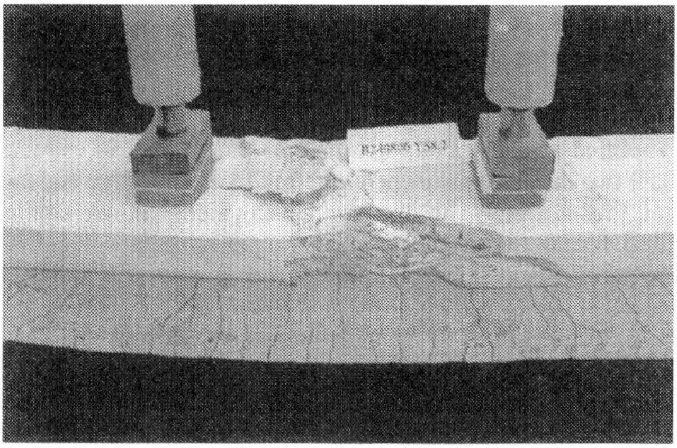

Figure 3. Reinforced concrete – the combined material of concrete and iron – is a genius combination. Concrete takes up the compressive stresses, iron the tensile stresses, and concrete and iron work well together: There is good adherence between concrete and iron, they have almost identical thermal expansion and the steel is passivated and protected from corrosion by the concrete.

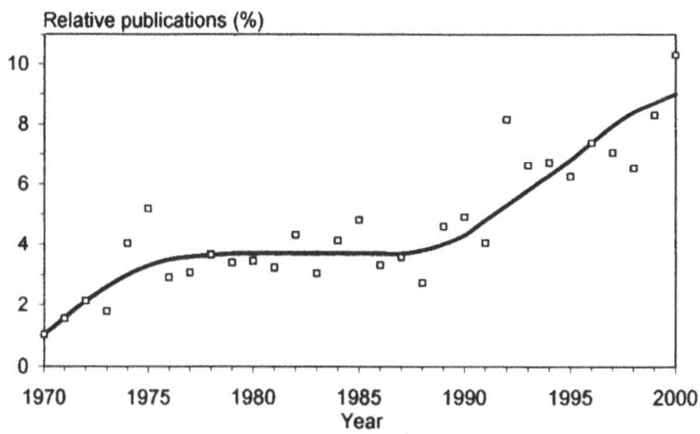

Figure 4. Percentage of publications in the scientific cement/concrete literature, which deals with fiber reinforcement. The curve indicates the renewed interest in fiber reinforcement, which was initiated around 1970 and 1990. The curve has been constructed with the literature database Compendex.

Different reinforcement technologies are combined in compact reinforced composite, CRC, which was developed and patented by Bache in 1986[9,10]. CRC is an ultra-high-strength concrete reinforced by 5-10 vol-% steel fibers together with 10-15 vol-% conventional reinforcement or prestressed reinforcement. CRC is a new material where a new structural design philosophy has to be used. Properly designed, structures of CRC can be made with the same load carrying capacity and the same material volume as steel, but with half the weight. Strength and ductility are significantly higher than for conventional concrete and resemble the properties of steel.

CHARACTERISTICS OF DEVELOPMENT

As exemplified above, technological development may in extreme cases occur through two different paths, here referred to as deductive and inductive development, respectively. Deductive development is a continuously expanding and directly logical progress based on existing knowledge. As such, deductive development is global and prescriptive in content. Contrary to this, inductive development is particularistic and descriptive in content. Inductive development is characterized by discrete, innovative advance that logically is only indirectly connected to existing knowledge. Inductive and deductive development may be considered nucleation and growth of knowledge, respectively.

Effective development requires both deductive and inductive progress since they - so to speak - are complementary. Deductive development will meet barriers that are easiest to overcome through inductive development. On the other hand, inductive development needs to be utilized by coherence and growth from deductive development. These two opposite types of development are in many aspects analogous to the human thinking process and may have a biological explanation as emphasized by de Bono[11,12].

Figure 5 illustrates the above points. Examples of inductive development within reinforced concrete are fiber reinforcement and prestressed reinforcement. Examples of deductive development include changes to fiber materials and optimization of fiber geometry based on fracture mechanics. Different technologies may eventually coalesce due to combined inductive and deductive development. CRC is an example of this.

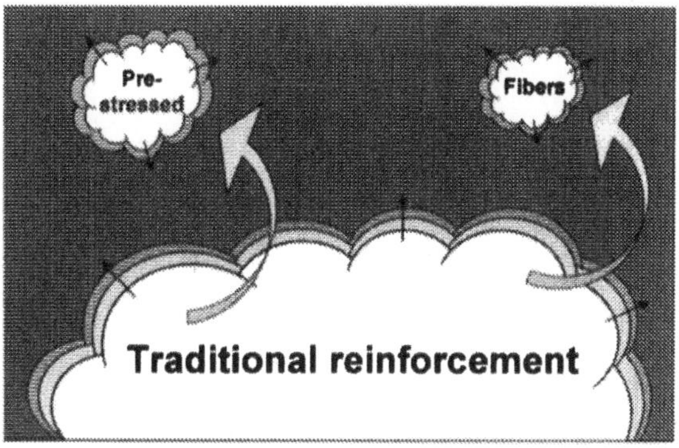

Figure 5. Technological development within concrete reinforcement has taken place as a nucleation and growth process: through discrete steps and by continuous progress, referred to as inductive and deductive development, respectively. Inductive development is shown by large, curved arrows, and deductive development by small, straight arrows.

Technological change often passes through certain stages in a characteristic manner. Initially a concept is nucleated, and if it is sustainable it may develop and spread. During this growth, the concept is optimized and rationalized deductively. Eventually, a limit for the technological development of the concept is approached, leading to stagnation. A new concept may finally appear and substitute or erode the existing concept. Also, a change in market needs may cause the erosion[13,14]. This characteristic development may be found within both products and processes, as well as in knowledge build-up.

Figure 6 shows the resulting typical S-shaped (sigmoid) growth curve. The curve shape reflects the lapse of time between nucleation and steady growth, and the relatively long period of time over which the concept is gradually adopted by all of the relevant agents. This simplistic form does not always apply. On a larger scale, technological development may consist of many separate steps as indicated in Figure 5. This may lead to a more complex growth curve (see Figure 4).

A peculiarity of the progress within concrete technology is the long time it takes from when a new concept appears until most potential users have adopted it[15]. A period of fifty years is not unusual – see, for example, the above description of the development of reinforcement technologies. One reason for this slow growth is the general conservatism within the building technology. Codes, standard specifications and test methods are elements in this.

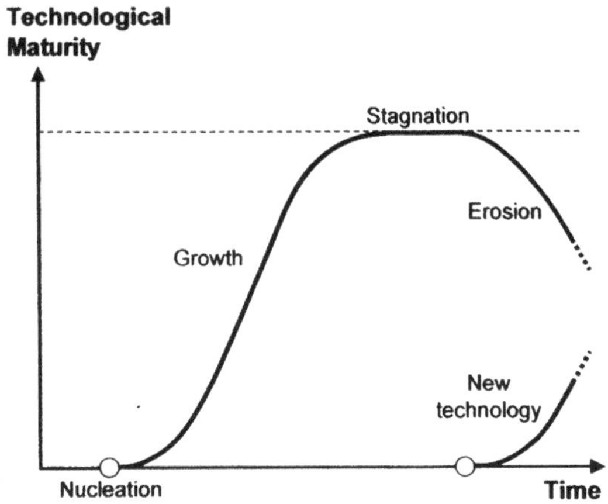

Figure 6. The growth curve of technological development often has a characteristic S-shape. Between nucleation of a new concept until it is fully developed and utilized, a very long time may pass.

An underlying barrier to effective development within the concrete field is the high degree of fragmentation of the concrete enterprise. The concrete industry is composed of many small producers and users and there is a low degree of cooperation between them[15,16,17]. Effective use of prestressed concrete, fiber reinforced concrete and CRC, for example, presupposes an associated, new design philosophy. However, this requires that architects and structural engineers understand and appreciate the qualities of these inventions. Generally, a new, inductive development is only of direct benefit for the particular company that invented it. The necessary volume growth and coherence of the development do not occur when other companies lack incentives in its promotion.

A further complication specific to the concrete field is the difficulties in practical utilization of new knowledge. This will be addressed below.

THE RESEARCH – PRACTICE SCHISM

In the past 50 years, an electronics revolution has taken place. Tremendous advancements in the scientific understanding of the physics of metals, dielectrics, and semiconductors has taken place. This, together with processing techniques such as thin-film growth and deposition, and photolithography, have allowed the development and production of transistors, microprocessors and high-density memory[18]. Within the electronics field, a very large part of the increased scientific knowledge has been applied directly to practice; even in practice it is possible to control photons and manipulate electrons. Most sectors of the manufacturing industry have undergone a similar technological revolution in materials and processes.

Today frontline concrete research is also a highly advanced discipline. For example, the hydration process is analyzed by complex computer simulations, and chloride ingress and chloride binding are followed by Electron Probe Micro Analysis and Nuclear Magnetic Resonance. However, these developments have not been paralleled within concrete technology, and have been even less successful in concrete practice. Specific parts of the concrete industry have developed considerably, such as the prefabrication industry, the ready-mix industry and the concrete product industry, but despite this the concrete industry is still relatively low-tech, low-skill and labor-intensive. Many problems in concrete practice still seem to be awaiting scientific treatment. The labor-intensive curing safeguards are one example of a problem that may be solved scientifically. Self-compacting concrete is an example of a recent, scientifically based development that has made concreting less labor-intensive.

Figure 7 illustrates this schism. Despite substantial scientific knowledge in the concrete field, only a small amount of this is applicable in practice[4]. Transfer of knowledge, rather than the amount of knowledge, seems to be the limiting factor for concrete progress. The adaptation of science (i.e. technology) is therefore vital in the concrete field, contrary to other fields, where new theoretical and experimental knowledge may be utilized directly in practice.

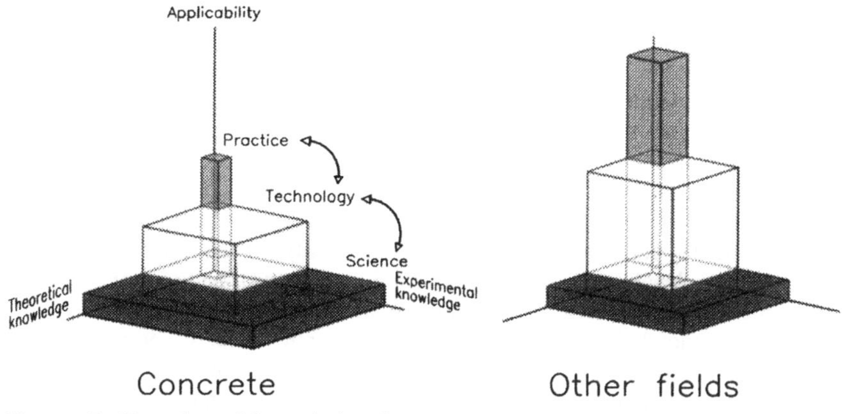

Figure 7. Transfer of knowledge between concrete science and concrete practice is inadequate. Only a small amount of the scientifically produced knowledge is utilized in practice compared with other fields, e.g. electronics. Science is here defined as the production of knowledge, technology as the adaptation of science, and practice as the utilization of technology. Based on Olesen et al[19].

The underlying reason may be that the materials used in electrical and mechanical engineering are, to a high extent, well defined. In addition, their required material properties are also well defined. Contrary to this, concrete is traditionally an inhomogeneous, cheap and large-volume product, which has a complex interaction with other building materials. Furthermore, concreting often has to take place under the influence of nature. Effective industrial development of the building industry is prevented since most buildings are unique products. As a result, it is very difficult to transfer new scientific insight into practical application; the route from the concrete research and development laboratory to the reality of the building site is stony.

PRESENT AND FUTURE CHALLENGES

It is easy to identify challenges in concrete technology. Areas with major potential for improvements include labor and energy consumption. Concreting generally implies several labor-intensive processes, e.g. insurance of proper curing and construction of formwork. Internationally, the environmental aspect is becoming more and more important. The wish for sustainable development results in environmental considerations leading to, for example, recycling of concrete and more focus on durability. This tendency is most distinct in the highly industrialized countries, but is also gradually being acknowledged in the rest of the world. Fundamentally, concrete is an environmentally friendly material. Basic concrete materials are born out of nature and concrete and cement production may even turn waste products, e.g. fly ash, to good account. However, since cement is an energy intensive material it is alluring to search for ways to utilize cement more effectively and to produce more energy-effective cements. Especially for high-performance concrete, there is a potential for a more effective use of cement; the ultimate degree of hydration of cement may not exceed 50% in high-performance concretes. This means that half of the cement is used merely as inert filler rather than as the binding agent it is produced to be[20].

The future for concrete seems bright. No material on the technological horizon seems to be able to take significant shares of the concrete market. Concrete has continuously expanded its versatility for the last 150 years, and innovative developments will surely continue to appear and improve concrete in the future. An underlying challenge of the present and the future is to improve knowledge transfer within the concrete field: to bring scientific insight into practical application at the construction site – and to identify problems and possibilities at the construction site and give feedback to the scientific research laboratory. This would also promote a change towards development of materials technology based on advances in fundamental science, rather than the prevailing emphasis on empirical engineering and workmanship that characterizes the building and construction industries.

REFERENCES

[1]M.F. Ashby, "Materials selection in mechanical design", 2nd ed., Butterworth-Heinemann, Oxford, 2000

[2]P.-C. Aïtcin, "High-Performance Concrete", E & FN SPON, London, 1998.

[3]J.L.S. Aspdin, "Artificial stone", Patent No. 5022, London, 1824.

[4]H.-H. Gotfredsen and G.M. Idorn, "Curing technology at the Faroe bridge", Denmark, ACI SP 95-2, 1995, p. 17-31.

[5]J.E. Gordon, "The New Science of Strong Materials", 2nd ed., Penguin Books, Harmondsworth, 1986.

[6]E. Suenson, "Reinforced concrete" (in Danish), 1st part, 2nd ed., P.E. Bluhmes boghandel, Copenhagen, 1918.

[7]R.F. Zollo, "Fiber-reinforced Concrete: an Overview after 30 Years of Development", *Cement and Concrete Composites*, **19** [2] 107-122 (1997).

[8]ACI committee 544, "State-of-the-Art report 544.1R-82 on Fiber Reinforced Concrete", *Concrete International*, **4** [5] 9-30 (1982).

[9]H.H. Bache, "Compact reinforced composite article - with reinforcement embedded in fibre-reinforced matrix", Patent WO8707597-A, 1987.

[10]A. Loukili, A. Khelidj and P. Richard, "Hydration kinetics, change of relative humidity, and autogenous shrinkage of ultra-high-strength concrete", *Cement and Concrete Research*, **29** [4] 577-584 (1999).

[11]E. de Bono, "Lateral thinking", Penguin, Harmondsworth, 1971.

[12]E. de Bono, "The mechanism of mind", Penguin, Harmondsworth, 1971.

[13]J.C. Fisher and R.H. Pry, "A simple substitution model of technological change", *Technological forecasting and social change*, **3** 75-88 (1971).

[14]F. Lissoni and J.S. Metcalfe, "Diffusion of innovation ancient and modern: A review of the main themes"; pp. 106-141 in *The Handbook of Industrial Innovation*, Edited by M. Dodgson and R. Rothwell, Edward Elgar, Cheltenham, 1996.

[15]G.M. Idorn, "Concrete Progress – From antiquity to the third millennium", Thomas Telford, London, 1997.

[16]Committee on Status of Cement and Concrete R&D in the United States, "The Status of Cement Concrete R&D in the United States", National Materials Advisory Board, NMAB-361, 1980.

[17]W.H. Plenge, "Vision 2030: A Vision for the U.S. Concrete Industry - Our industry's 30-year map to meet future goals", Concrete International, **23** [3] 25-34 (2001).

[18]A. Polman and P. Wiltzius, "Materials Science Aspects of Photonic Crystals", MRS Bulletin, **26** [8] 608-610 (2001).

[19]S.Ø. Olesen, P.F. Hansen and K.R. Henriksen, "Concrete Curing", *Dansk Beton*, **2** 81-90 (1990).

[20]G.M. Idorn, "Cement use innovations: Necessary and possible", *Materials and Society*, **1** 119-123 (1977).

Photo 9: O.M. Jensen

Photo 10: Å. Skarendahl, D. Mitchell, K. Rear and N. Thaulow

DEVELOPMENT OF NEW GENERATIONS OF CONCRETES IN POLAND

Andrzej M. Brandt
Polish Academy of Sciences
Institute of Fundamental Technological Research
Swietokrzyska 21
Warsaw, Poland, 00-049

ABSTRACT

During the long period of the Soviet Union's influence in Poland after WWII (1944-1989) large delays in development were accumulated in many fields. Now, twelve years later, considerable progress has been achieved in research and technology in the field of concrete and concrete structures.

Bases for reconstruction and development of the civil infrastructure have been laid down. There are several research centers at universities and at institutes that are working in concrete and concrete structures. The number of students at universities has increased by a factor of 3.5. However, there are still several barriers to overcome; some of them are particular to Poland, others are similar to those in other countries:

- financial support for research from the government is insufficient.
- construction companies are relatively weak and cannot sponsor long-term research.
- transfer of new technologies to practice is rather slow or based mostly on imported techniques.
- foreign companies working in the country use almost exclusively technologies developed in their laboratories abroad.

The urgent needs for development stimulate large programs in building and civil engineering (apartment buildings, highways, railways, airports, etc.) and that is the foundation of moderate optimism. New trends adopted by advanced countries, like sustainability and friendliness of construction to the ecological systems, should be further applied and developed. This requires intensive research and international cooperation in the field of concrete and concrete structures.

PRODUCTION OF PORTLAND CEMENT IN POLAND

The first Portland cement plant built in Poland was in 1857 in Grodziec, the 5th such plant in the world. Its production was 490 tons/year. By 1939, production in that modernized plant was 360,000 tons/year and total production in Poland 1.7 million tons per year.

Figure 1: First Portland cement plant in Poland, fifth in the World – Grodziec 1857. *(Polski Cement, with permission).*

After WWII, cement production increased steadily and in 1999 was 16.0 million tons per year. In 2000, the dry process was used in approximately 70% of cement production, with low energy consumption and low emission of gases and dust. At present, over 30 different kinds and classes of cement are produced.

RECONSTRUCTION OF THE INFRASTRUCTURE IN POLAND

The primary task in building and civil engineering is renewal of the old and inadequate infrastructure system, i.e. construction of:

- new highways and motor ways
- high-speed railways
- several bridges of medium and long spans
- underground railway system in Warsaw; and other big towns.

All of these works require large volumes of good and durable concrete, sometimes also special concretes, like HPC and FRC.

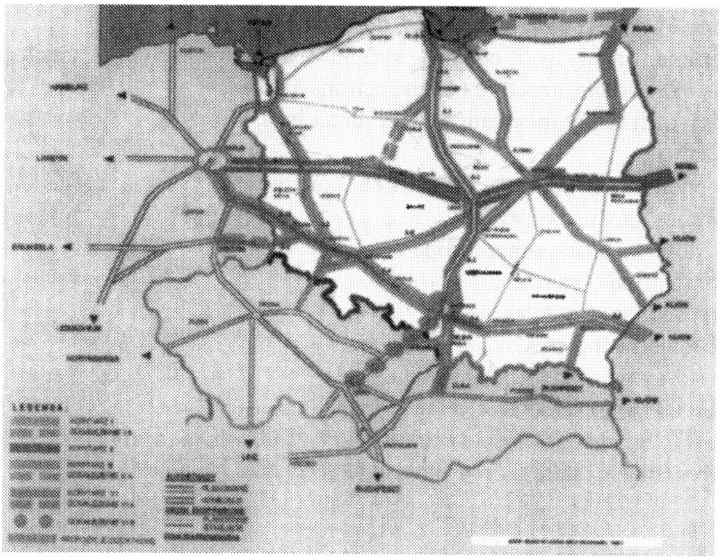

Figure 2: System of highways and motorways under construction in Poland.

RESEARCH IN BUILDING AND CIVIL ENGINEERING

Research is realized in three basic sectors: universities, institutes belonging to various ministers, and institutes of the Polish Academy of Science. Funding for research in general, and also in building and civil engineering, is insufficient. Additional funds are provided from international grants and joint projects. The involvement of Polish centers in international research is slowly increasing.

Participation of the building industry in research is limited, because building companies are small and not really interested in the sponsoring of any serious research. Their interest is concentrated on emergency situations: when cracks appear they ask for help and possible advice, sometimes related to a program of testing. Preliminary consultations to avoid possible damage are infrequent. Big companies have their centers abroad, e.g. cement industry. Therefore, new products and technologies are frequently prepared abroad and brought to the country without participation of local research centers.

The main directions of research in concrete are:

- durability (freezing, corrosion of steel, carbonation, fatigue, chlorides, sulphates, realkalisation),
- composition of the mix (superplasticizers, silica fume, fly ash, etc.),

- application of ANN (design of mixes, durability),
- computer image analysis of concrete structure, air-entrained pores, cracks,
- diagnosis of concrete structure (plain sections analysis),
- application of fracture mechanics (FRC, cracking),
- curing methods,
- precast concrete elements, curing, cellular concrete,
- re-use of concrete, use of wastes,
- strength, deformability, other properties,
- polymer modified concretes,
- high performance concretes
- fiber reinforced concretes

EXAMPLE OF SPONSORED RESEARCH

NATO Science for Peace Project 1999 – 2003 "Diagnosis of Concretes and High Performance Concretes" by Structural Analysis".

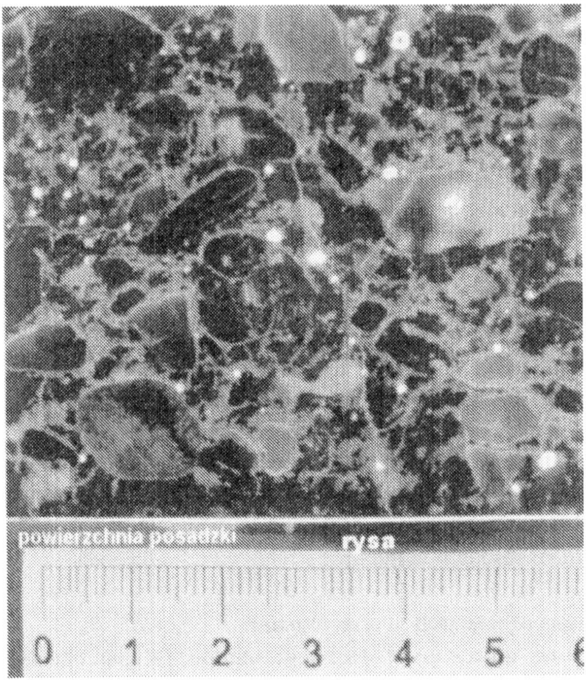

Figure 3: Image of concrete structure from a plain section of a specimen, prepared for quantitative analysis.

Applied methods:

- micro- and macro-images of plain sections of specimens and cores
- polishing and impregnation
- computer image analysis of the structure
- quantitative determination of objects (pores, cracks, aggregate grains, etc.)
- mechanical testing (strength, strain, Young modulus, etc.) and indentation
- acoustic emission recording and analysis
- application of Artificial Neural Networks and other soft methods for analysis of results

Partners are research laboratories from Denmark, Belgium, US and the Institute of Fundamental Technological Research in Warsaw.

The aim is to improve the quality of infrastructure through the application of new methods of diagnosis of concrete with respect to its strength and durability against freezing and thawing cycles.

DURABILITY

Durability of concrete structures is one of the most important directions of research. The requirement of a durable structure is met if, throughout its required life, a structure fulfils its intended function with respect to serviceability, strength and stability without significant loss of utility or excessive unforeseen maintenance.

ORGANIZATION OF A SERIES OF INTERNATIONAL SYMPOSIA ON BRITTLE MATRIX COMPOSITES

1 – 1985 Jablonna
2 – 1988 Kielce
3 – 1991 Warsaw
4 – 1994 Warsaw
5 – 1997 Warsaw
6 – 2000 Warsaw
7 – 2003 Warsaw, October 15-17

For each Symposium, a volume of proceedings is published. The papers are presented, discussed and published only in English. On average, there are around 100 participants in each symposium, representing several countries.

Figure 4: Warsaw Financial Center Bldg. reinforced concrete structure, height 144 m, pumped concrete 150 m, $f_c > 65$ MPa, built in 1997-1998.

Figure 5: Prestressed concrete bridge over Vistula in Toruń, main spans 3x130 m, length 995 m, built in 1998.

Figure 6: Two prestressed concrete viaducts in Świnna Poręba, length 170 and 217 m, built in 1999.

Figure 7: Prestressed concrete bridge over Odra in Opole, main span 100 m, length 385 m, built in 1999.

Figure 8: Suspended bridge over Vistula in Warsaw, main span 250 m, total length 500 m, under construction 2000-2002 (reinforced concrete pylon). *(Polski Cement, with permission)*.

CONCLUSIONS

- There is a large need for development of concrete structures in Poland.
- The quality of concrete and mainly its durability should be improved.
- The research potential of Polish laboratories is increasing rapidly to cover existing delays and to fulfill requirements of the building industry.
- There is a strong need for collaboration in research with the laboratories in the most technically advanced countries.

Photo 11: A. Brandt

Photo 12: O.M. Jensen, D. Myers and S. Diamond

CONCRETE TECHNOLOGY IN FORMER EAST EUROPEAN COUNTRIES: CHALLENGES & OPPORTUNITIES

Adolf Bajza and Igor Hudoba
Faculty of Civil Engineering
Slovak University of Technology,
Radlinského 11, 813 68 Bratislava,
Slovakia

ABSTRACT

The transformation process in higher education, research and science in Slovakia and some other post-communist European countries after 1989 is discussed. Information about changes in the construction industry and in concrete technology, and challenges for the future, are also presented.

INTRODUCTION

November 1989 was the beginning of the transformation of Slovak society as well as higher education. Slovakia became an independent and sovereign state on January 1st, 1993, following the separation of the former Czech and Slovak Federative Republics, and expressed the desire to become a new member state of NATO and a member country of the European Union. The majority of the citizenship is oriented toward, and supportive of, European integration. In can be said that Slovak society is relatively unified from the point of view of the general principles of democracy and, in comparison with other post-communist countries, there are no remarkable differences. The free-market system has been accepted by the majority of our population. There are, of course, numerous problems that we are trying to solve at the political level. The whole new political system needs a certain amount of time to mature. In Slovakia, democracy is young and we all have to learn how to maintain it and support it in our daily lives.

Since 1989, the universities and higher education institutions have been playing an important role in the process of stabilizing the Slovak society and its democratic system. We are aware that education represents an important force in the process of economic, social, and also political transformation of Slovakia.

In the construction industry, the dramatic political changes of the 1990s led to the interruption of large housing projects and to changes in the ownership of construction companies. Funds necessary for new infrastructure and housing were unavailable. This situation also dramatically affected the ready-mixed concrete industry.

HIGHER EDUCATION IN SLOVAKIA DURING THE LAST TWELVE YEARS

The adoption of the New Act of Higher Education in Slovakia, which was approved by the parliament and become valid in June 1990, was the crossroads or the milestone in the modern history of Slovak higher education. This new act has opened the windows and allowed fresh breezes to enter the universities' halls and freshened the entire Slovak academic society. Official communist ideology and the leading role of the communist party were eliminated.

The basic principles of university autonomy, academic freedom and rights were codified in the new act. Academic senates, self-governing bodies playing the role of an academic parliament at the university and faculty levels, were created. Widespread and vital international collaboration has developed practically at all universities and other higher education institutions. The process of comparing different curricula and study schemes, as well as new curricula development, has begun. The academic society has adopted a system of two diploma levels: the bachelor and the master level. In another words, the European format was incorporated into Slovak higher education.

One of the most important recommendations addressed in 1993 was to initiate the process of diversification and differentiation of higher education. Another important issue related to the expected increase in the percentage of 18 year olds pursuing higher education, from the present 16% to an expected 30% in the near future. Looking back with present day eyes, we can evaluate the situation and we are able to say what were the positive and the negative changes in our higher education in the last twelve years.

The existing Act of Higher Education contains one very significant specific item; that a faculty, as a university unit, is a separate legal entity. All faculties within one university have the full power to make decisions in all areas of a given faculty's activities, including economic and financial issues. Each faculty is able to organize its own educational and research activities independently from other faculties and university units. Each faculty has its own department of mathematics, physics, foreign languages, etc. Therefore, there are numerous duplications within one university. Such a system cannot be effective. In such an environment, it is extremely hard to manage the university as a single institution. In contradiction to the present situation within universities, and from the point of view of efficiency, the general conditions outside the university are worsening from year to year. Whereas public funds allocated by the state to academia continuously decrease, universities are extremely interested in ensuring quality of education and research, as well as in contributing to the development of knowledge and science. This desired contribution applies not only to the national level, but also to world-wide competition and collaboration in most fields, including the humanities and social sciences.

The present system or model of higher education in Slovakia can be characterized as a unitary system. According to valid laws applicable to Slovak higher education, all of the higher education institutions have been given "the status of university", which essentially means "research university". All Slovak universities have the same rights, the same duties, practically the same methods of financing, and use the same processes of evaluation and accreditation.

In general, higher education in Slovakia is facing the permanent problems of financing. In comparison with other European countries, Slovakia ranks 22nd out of 24. Slovakia provides only 0.67% of its GDP to education! In comparison, for example, Norway allocates 2.16% of its GDP to higher education. There is also a long-term internal debt, representing more than five billion Ks, which is needed for maintenance, repair and reconstruction of existing university buildings and their basic equipment.

One of the most serious problems facing Slovak higher education is the age of teaching staff. The average age of full professors is over 62 and of associate professors over 57. The teaching staff at Slovak universities has become an aged society. The category of young teachers and researchers between the ages of 25-35 is virtually non-existent in some faculties and departments.

These conditions result in week motivation of young graduates when deciding upon their future academic careers at the university. There exists a brain-drain from universities to the private sector and abroad, where the conditions are much more lucrative for young university graduates.

PRESENT STRUCTURE OF HIGHER EDUCATION IN SLOVAKIA

In Slovakia there are in total 21 higher education institutions that are state organized, state-run, and state-financed. At the present time, there is only one private higher education institution in Slovakia, the City University of Trencin, a branch of the City University, Bellevue, USA, offering an MBA diploma.

Altogether, there are currently over 110,000 students in higher education in Slovakia.

RESEARCH AND SCIENCE IN SLOVAKIA

In Slovakia, research and science activities are concentrated in three areas: universities, the Slovak Academy of Sciences, and industrial research institutions. The research activities of the Slovak Academy of Sciences were oriented mostly toward basic research in the past. According to the existing law, all 21 higher education institutions must pursue research activities, with an Accreditation Board approving the quality of research. This body approves all courses at the B.S., Mgr., and Ph.D. levels. On the other hand, there are big differences in the number of Ph.D. students among the existing universities. The two leading universities, Comenius University and the Slovak University of Technology, both located in Bratislava, educate over 60% of all Ph.D. students in Slovakia.

At present, there are four practical ways of financing research activities:

1. Institutional financing as part of the university budget allocated by the state.
2. VEGA - the national research agency provides state money via proposal competition.
3. Direct contract with industry.
4. Co-operation in international research programs, international research grants.

In recent years, only about 0.7% of the Slovakian GDP has been allocated for research and science. Compared with the average percentage in neighboring countries and with countries of the European Union, this is insufficient.

An Act of Financing of Research and Technologies and an Act of Academy of Sciences are currently under development. The future of the Slovak Academy of Sciences, as the second most important center for research and science in Slovakia, is unresolved.

EXPECTED DEVELOPMENTS IN HIGHER EDUCATION IN SLOVAKIA

In spite of the many above-mentioned problems, the Slovak universities have a good reputation. We have been able to transpose the new situation in society as a whole into the academic environment. The continuous process of new curricula development offers our graduates the opportunity to be well prepared for the domestic and international labor markets. Nowadays, the knowledge of foreign languages is completely different when compared to the situation twelve years ago. Isolationism toward information from abroad was at that time a part of state policy.

Very recently, a New Concept of Future Development of Higher Education in Slovakia was prepared and introduced[1]. This new concept includes the basic principles of European integration and international compatibility[2,3]. The following issues are considered to be the most important:

- Increasing the number of students and their access to higher education,
- Creating partnerships with economic life,
- Developing other forms of education (continuing education, open and distance forms of learning),
- Adopting the European formats in higher education,
- Participating intensively in the European programs of cooperation in education, research and science (Copernicus, SOCRATES, LEONARDO, 5th Frame Program of Collaboration in Research and Science).

The term "European formats" relates mainly to the mobility of students and teaching staff, international cooperation, common research programs, recognition of qualifications and diplomas, etc. The basic precondition for the transformation of the existing system of higher education in Slovakia and its integration into the European system is its institutional, content, and quality diversification. It means that the university status of some existing higher education institutions has to be changed to non-university status, serving especially market-oriented industries.

Slovakia adopted the so-called "Bologna Declaration" in June of 1999 and the three-level system of B.S., Mgr., and Ph.D. degrees has been adopted.

In the frame of the New Concept of Future Development of Higher Education in Slovakia a detailed analysis of the existing model was realized. Both the strong and the weak points of the existing model were determined.

The strong points are:

- Real autonomy and academic freedom.
- Great potential for students; a large number of talented young people who are interested in studying at university.
- Large selection of enthusiastic and creative people at the universities.
- Large number of highly qualified people outside higher education who are able to collaborate with the universities.
- Good quality of graduates leaving Slovak universities; ability to find employment at domestic and at European labor markets.
- Low rate of university graduate unemployment (less then 5%).
- Ability of higher education institutions in Slovakia to join international collaborations on education and research.
- Large scientific and research potential at the Slovak universities waiting for application.

The weak points are:

- General concept of the whole system of education in Slovakia is missing.
- Mission of the individual higher education institutions is missing.
- Diversification of higher education is missing (institutional and program oriented).
- Lack of finances; present single financing method is insufficient, need for multi-source system of financing.
- Lack of young teachers and researchers at the universities.
- General moral and ethical crisis in society and its reflection on higher education.
- General conditions for education and research activities at the universities get worse from year to year.
- Graduates from Slovak universities are overeducated.

- Internal organization of the universities is inflexible; there is minimal internal student mobility, and insufficient university management because of incorrectly divided competencies.
- Bad situation in the university libraries.
- Outdated information technologies.
- Infeasible, complicated, and inconsistent system of diploma.
- Too many study courses and subjects of study.
- Insufficient level of quality in university management; democratically elected academics are acting in positions of management professionals (for three, maximum six years).

The New Concept of Future Development of Higher Education in Slovakia[1] was the basis for preparation of the proposal of the New Act of Higher Education[4]. This act is under discussion and is expected to be passed in the National Council of the Slovak Republic (National Parliament) in a few months. The New Act should become law by January 1st, 2002.

The New Act of Higher Education will create a fully new environment at the universities. The most important changes, compared to the existing law, are expected to be:

- State universities and higher education institutions will change their status to public.
- The present uniform model of higher education will be diversified into university and non-university sectors.
- The existing model for single-source financing of higher education institutions will be replaced by a multi-source model.
- Mobility of students and teachers will be enhanced; credit system, evaluation, and accreditation according to international standards will be adopted.
- Different forms of organization and financing of research at the universities will be adopted.

TRANSFORMATION PROCESS OF CONSTRUCTION INDUSTRY, CHALLENGES FOR FUTURE

The last twelve years were a period of radical changes for the construction industry in Slovakia. The change in socio-economic relations, the fall of the old structure of the state and cooperative enterprises and contracting companies, and the process of privatization brought with them numerous problems. Large state-run companies have mostly disappeared, and a new structure of small and medium size private companies and firms appeared. Only a few large companies operating in the area of transportation (highways, road, and bridges) have survived.

In general, the construction industry in Slovakia depends primarily on state investment policy and public money. The amount of public finance, as a part of the state budget, depends on overall economic growth. From this point of view, the construction industry in Slovakia has been in a state of depression over the last few years.

The construction industry in Slovakia provides the largest proportion of jobs to the population and thus depression in the state economy is reflected immediately in this sector. In this respect, the construction industry can be taken as an indicator of economic growth. During the last twelve years, the number of employees in the construction industry has dropped by 50%. In comparison with 1989, the actual production efficiency decreased by 30%. The main reason for this is the declining demand for building activities, since the construction industry was, until now, not fully prepared for the free market economy. State policy in the areas of taxes, bank credits and employment is creating low motivation for private building contractors.

It is not only the environment in the production area of the construction industry that has been fully changed. The area of construction design has also been transformed. Former large state-run design institutes have disappeared. Design activities for the construction industry are now managed by small private design offices. The Chamber of Civil Engineers is the official body that guarantees the quality of design and production in the Slovak construction industry.

One of the very visible positive changes in the construction industry in Slovakia is the emphasis on quality. It is the natural reflection of the new competitive environment and the elimination of state limitations and price regulations, which were so typical for a centrally planned economy. Nowadays, the quality of building and construction is fully compatible with the quality abroad.

Because of the above-mentioned depression in the construction industry, research activities in the area of civil engineering are significantly reduced. Also, acceptance of research proposals in domestic and international competitions in the field of civil engineering is low.

CONCRETE AND CONCRETE TECHNOLOGY IN SLOVAKIA AND OTHER CENTRAL AND EASTERN EUROPEAN COUNTRIES

Concrete, as a construction material, has been playing a decisive role in the construction industry during the last half century in Slovakia. There are enormous natural sources of excellent quality aggregates, mostly in the western part along the Danube river. This reservoir is considered to be one of the biggest in Europe.

The situation in the cement industry has changed too. Many foreign investments in cement production were realized in last twelve years (e.g. Holcim, Lafarge, etc.). The quality of cement made in Slovak cement factories is compatible with the European standard. A large portion of the cement production is exported.

Prior to 1989, ready-mixed concrete production was concentrated in a few RMC plants situated in the vicinity of large construction sites, mainly those for housing construction. They were organized in cooperation with construction companies established for large construction projects planned by the government and did not serve like in Western Europe[5]. Usually, it was the General Manager himself who decided which construction site receives what and when. It was not a market situation at all. Every construction company made its own statistics. Separate data on the production of RMC were never available.

After 1989, the construction industry stopped receiving large projects, mainly due to the fact that the governments of CEEC were unable to find funds for housing and infrastructure projects. The question was how to survive the new situation without markets, projects, and jobs. The largest construction enterprises split into smaller parts and were then privatized. One such part was usually the concrete production sector of the company. The concrete production plants, truck mixers and pumps were separated, and then sold to new owners. Very often, joint ventures were formed - one partner from the original construction company and the other one usually with financial support from abroad.

In a relatively short time, RMC producers from the Czech Republic, Slovakia, Poland, Hungary, Croatia and later Rumania and Bulgaria became parts of RMC companies from Western Europe. This transition process was very difficult. Some factors that played an important part in the survival of the new RMC companies can be summarized as[5]:

- Opposed interests of the partners in the joint-ventures (construction companies interested in securing low cost concrete, foreign partners trying to establish a professional RMC industry),
- Excess personnel working in concrete production with relatively low efficiency,
- Small and relatively old truck mixers,
- No funds for equipment repair or for the RMC industry,
- Non-existing market information,
- Black market, especially in big cities,
- Difficult financial situation caused by non-solvent customers receiving large amounts of concrete,
- Quality of aggregate, cement and admixtures.

Producers of RMC have found the solution to the problem in the foundation of RMC associations. Today, practically every CEEC has its own RMC association. They try to be associated with the Western European associations and cooperate with the European Ready-Mixed Concrete Organization (ERMCO). The Czech Republic and Poland are full ERMCO members. Slovakia, Hungary and Russia are corresponding ERMCO members at this time.

From the RMC point of view, the Central and Eastern European countries can be classified into two large groups[5]:

- Group 1 - Czech Republic, Slovak Republic, Poland, Hungary, Croatia and Slovenia with fully market oriented RMC industry, which is privatized. It is hoped that in the near future they will become members of the European Community. Some of them are also members of European Committee of Normalization (CEN). The
- majority of RMC producers are members of national RMC associations. There is therein a platform for the exchange of technical and economic information, for lobbying and for improving the image of the RMC industry.
- Group 2 - countries like Bulgaria, Rumania, Russia, Commonwealth of Independent States, and Balkan countries in which the RMC industry is going to be privatized. Many concrete producers are still associated with the large construction companies.

It is difficult to obtain full information about the concrete industry situation in CEEC. However, based on data in Tables I – V[6,7], it is possible to gain some understanding of the RMC industry from the point of view of producers, plants, transportation, consumption, employees and quality of RMC produced in some CEEC and Western countries.

Table Ia. Development of RMC Industry in Some CEEC - Year 1995[6]

Country	Producers	Plants	Truck mixers	RMC production Thousands m^3	Cement to RMC, %	Turnover Millions
Czech Republic	200	400	1800	5400	38	35000 Kc
Hungary	9	1	20	60	11,7	260,5 HUF
Poland	-	-	-	-	-	-
Russia	300	600	5500	27500	-	1650 US $
Slovakia	40	60	5	1200	-	28000 Ks

Table Ib. Development of RMC Industry in Some CEEC - Year 1998[6]

Country	Producers	Plants	Truck mixers	RMC production Thousands m^3	Cement to RMC, %	Turnover Millions
Czech Republic	200	400	1800	3900	-	25000 Kc
Hungary	9	1	20	60	39	260,5 HUF
Poland	250	350	1000	5000	-	1000 US $
Russia	320	800	5500	25100	20	1952 US $
Slovakia	60	95	290	2250	40	28950 Ks

Table II. RMC Industry - Year 1999[6]

Country	Plants	Production Millions m^3	Consumption m^3/ capita	Cement used kg/m^3	Coutries' employees
Austria	315	10.5	1.3	260	2400
Cuba	40	0.34	0.03	320	497
Czech Republic	400	3.9	0.38	310	15000
France	1650	33.2	0.52	270	6700
Germany	2351	62.3	0.76	290	14000
Hungary	1	0.06	0.00	309	28
Netherlands	185	8.00	0.50	333	2200
Poland	350				
Russia	1100	31.0	0.25	350	30000
Slovakia	100	1.8	0.33	330	900
UK	1250	23.55	0.41	280	10000
USA	5500	266.0	1.0	270	80000

Table III. Statistics of RMC Transport - Year 1999[6]

Country	Own vehicles %	On behalf of concrete plants %	On behalf of third parties %	Average number of trips per truck mixer	Average distance travelled to site km
Austria	58	40	2	4.0	15
Cuba	75	15	0	3.5	12
Czech Republic	50	24	26	4.0	13
France	20	70	10	3.8	15
Germany	37	56	7	5.0	15
Hungary	84	3	13	4.0	25
Netherlands	95	3	2	3.5	36
Poland	80	10	10	5.0	20
Russia	5	80	15	4.5	15
Slovakia	35	60	5	3.6	8
UK	24	75	1	6.0	5
USA	98	1	1	4.0	20

It is also interesting to compare prices for 1 m^3 of RMC (1999) between some European countries: Austria 38.1€, Czech Republic 31.9€, Hungary 32.2€, Poland 44.6€, Slovakia 28.9€7. New technologies for mixing, transport and placing of concrete were implemented and are used in the RMC production in Slovakia. At present, there are twelve important private companies acting in transport-concrete production. They produce over 700,000 m^3 of ready-mix concrete per year. The total amount of RMC production in 1999 was over 1,800,000 m^3. It represents about 0.33 m^3 of RMC per capita[7]. Comparing this amount with the European average of about 0.80 m^3 per capita, it is less than one half[6].

Table IV. Production of Concrete and Consumption of Cement - Year 2000[6,7]

Country	Concrete production				Cement consumption				
	Total Mill. m^3	RMC %	On site mixed %	Precast %	Total Mill. tons	RMC %	On site mixed %	Precast %	Other uses %
Austria	14	66	10	24	4,6	52	9	24	15
Germany	105	55	10	35	35,75	47,1	9,4	31	12,5
Netherlands	15	53	2	45					
Portugal	25-28	40	35	25	11,1	24	20	12	44
Spain	124	52	23	25	38,5	42	18	20	20
Slovakia	3,26	57	28	15	1,63	40	20	10	30

Table V. Types of RMC Production - Year 2000[6]

Country	Strength class (1) [%]				Slump class [%]		
	< 15	15-C25	25-C35	> 35	S1	S2-S3	S4-S5 and above
Austria	10	35	40	15	5	45	50
Cuba	3	61	36	0	0	26	74
France		60				80	
Germany	4	13	62	21	11	84	5
Netherlands	2	71	24	3			
Russia	10	70	15	5	5	90	5
Slovakia	15	50	30	5	10	80	10
UK	18	35	39	8	3	96	1

(1)Strength classes according to ENV 2068

In spite of the present slow growth in the construction industry, the situation in RMC is characterized by increasing production. RMC production in Slovakia is becoming more and more sophisticated, and there is a clear tendency to offer full services to customers in the future.

The new situation in concrete technology and concrete industry was not fully reflected in practice until now. Because of bad past experiences and the typical "caution" of designers and civil engineers, there was a preference for lower quality (lower strength) concrete in construction design, despite the fact that concrete mix producers can routinely guarantee higher quality[6,7]. This is the reason why the most common concrete strength class was C16/20, a situation that has to be changed.

High strength and high performance concrete is the subject of research activities at the Faculty of Civil Engineering in the Slovak University of Technology in Bratislava. Unfortunately, there are only limited initial applications of high performance concrete in the Slovak construction industry. During the last years, however, closer cooperation with selected construction companies has been initiated.

CONCLUSIONS

New links between universities and the construction industry in Slovakia have been successfully developed. In spite of this progress, collaboration in the fields of applied research, technology and new materials has to be re-emphasized. Companies operating in the concrete business have to closely follow the general trends of future developments. However, these developments depend not only on the technical and economic ability of the partners involved, but new legislative tools, regulations and standards will have to be prepared and codified.

All Central and Eastern European Countries made substantial progress in the development of their concrete industries over the last twelve years. At the beginning of the transition period, there were many economic, political and institutional problems connected with dramatically decreasing concrete production in all countries. Nowadays, it seems that the bottom of the economic recession in the majority of CEEC has been reached and there are clear signs of growth in the construction industry as well as in concrete production.

The new competitive environment of the free market within Europe is a great opportunity for the concrete industry, but it is also a great challenge for all the players involved.

REFERENCES

[1] The Concept of Future Development of Higher Education in Slovakia until 2015, prepared by Commission of Experts, Ministry of Education in Slovak Republic, August 2000.

[2] Science for the Twenty First Century, A new Commitment, International Conference Budapest, Hungary, June 1999.

[3] UNESCO World Conference on Higher Education for 21st Century, Paris, France, October 1998.

[4] Draft of the New Act of Higher Education, prepared by Commission of Experts, Ministry of Education in Slovak Republic, February 2001.

[5] T. Dratva, "Development and Actual Situation of the Ready-Mixed Concrete Industry in Central and Eastern European Countries", ERMCO Congress 2001, Berlin, Germany, 13. - 16.6.2001.

[6] Draft Report of Members' Activities 2000-2001, ERMCO, Egham, UK, 2001.

[7] P. Kňaze, "RMC in Slovakia", ERMCO Board Meeting, Berlin, Germany, 13.6.2001.

[8] ENV 206:1989, Concrete - Performance, Production, Placing and Compliance Criteria.

Photo 13: A. Bajza

Photo 14: J. Beaudoin

Photo 15: L. Ko, J, Gebauer, S. Shah and D. Shaw

Photo 16: Å. Skarendahl and K. Rear

FUTURE NEEDS IN CEMENT AND CONCRETE: A NATIONAL HIGHWAY PERSPECTIVE

Stephen W. Forster
Leader, Concrete Pavement Team, HRDI-12
Office of Infrastructure R&D
Federal Highway Administration
McLean, Virginia, USA, 22101

ABSTRACT

There are a number of factors driving the current and future requirements for concrete as well as other paving materials. First, on a national basis, the primary mode of paving operations is switching from new construction to repair, rehabilitation and reconstruction. Second, because of user demands, most activities will take place under a time constraint due to the need to minimize the disruption of traffic. Cements and concretes will therefore have to set and gain strength faster to enable early opening. Third, concrete mixtures continue to become more complex. Paving concrete typically employs Portland cement plus at least one additional cementitious or pozzolanic material. This complexity will likely only increase, while expertise within the DOTs is decreasing due to staff reductions and retirements. The responsibility for developing mixture designs and proportions is being assigned more and more to the contractor by the DOTs. The trend is towards performance-related specifications for paving concrete, which takes some of the burden off the DOTs, and also gives the contractor added freedom in the mixture design process. Despite the added complexity, the behavior of the cement and other concrete components must be compatible and predictable when combined. Fourth, performance and durability of the concrete cannot be sacrificed because of factors one, two and three. Finally, both initial cost and life-cycle costs of concrete paving should be reduced in order for concrete to maintain or enhance its standing as a competitive paving material.

INTRODUCTION

The total public road mileage in the United States as of 1997 was nearly 4 million miles[1]. Of this, approximately 3.1 million miles is considered rural, the rest is urban. Approximately 75 percent of this total is locally owned, 20 percent is State owned, and 5 percent is federally owned. This translates to approximately 170,000 miles of federal roads. The Interstate System is only one percent of the total mileage, but accounts for approximately 20 percent of total travel. Of the total road mileage, approximately 55 percent is paved with either asphalt or concrete[2], with the split of the paved portion being approximately 90 percent asphalt, 10 percent concrete.

There are approximately 580,000 bridges in the country, with 51 percent locally owned, 47 percent State owned, and 2 percent federally owned.

INFRASTRUCTURE STATUS

On a national basis, the primary mode of paving operations is changing from new construction to repair, rehabilitation and reconstruction. Of the total road mileage noted in the introduction, nearly 60 percent had a pavement condition rating of fair or worse in 1998. On the bridge side, in 1998 nearly 30 percent of the bridges were rated as deficient, either because of a structural deficiency, or because they were functionally obsolete. If you will recall, construction of the Interstate System (and many associated roads) occurred for the most part in the late 50's, 60's and 70's. It stands to reason that these pavements, and many others constructed during this period of new construction, have met or exceeded their design life and are now in need of attention. It is for this reason that the primary paving activities are repair, rehabilitation and reconstruction. This prevalent mode of construction and the associated factors of user demand, concrete complexity, performance/durability and cost are driving the current and future requirements for pavements and paving materials. These factors are discussed further below.

USER DEMANDS

New construction has the advantage of taking place in a new right-of-way where no highway facility previously existed, and therefore there is no pressure to maintain travel capacity. On the other hand, in the repair, rehabilitation and reconstruction modes, most activities will take place under significant time and space constraints due to the need to minimize the disruption of traffic. In many instances the primary objective is to maintain the number of lanes and traffic speeds as much as possible during construction. This means that repair and rehabilitation operations must be confined, and as rapid as possible.

As a consequence, the cements and concretes used and their behavior, must be compatible with these compressed work zones and timetables. Cements and concretes will have to set and gain strength faster in order to enable early opening to traffic. In addition, once complete, the resulting pavement is expected to last longer, so that the road users are not disrupted again for a long time. In the words of a recent workshop on pavement renewal, the goal is to "get in, get out, and stay out"[3]. User demands and delays can't be ignored, because depending on how they are calculated, user delay costs can easily outstrip the construction costs of a project.

CONCRETE COMPLEXITY

Concrete mixtures are much more complex than they have been in the past. We are no longer looking at a simple three-part combination of cement, sand and stone, plus a little water to make it work, as our concrete formulation. Paving concrete will most likely also include fly ash or some other pozzolan, ground granulated blast furnace slag or other cementitious material, an air-entraining agent to provide freeze-thaw durability or improve workability, and perhaps other ingredients. Structural concrete used for bridge construction can be even more complex. In addition to the materials listed for paving concrete, bridge concrete will likely also include silica fume, water-reducing and/or plasticising admixtures, corrosion inhibitors, fibers, and perhaps other components.

At the same time as concrete complexity is increasing, control of the concrete properties and their development must continue to be improved in order to meet the time constraints often imposed during construction. We must control workability, setting time, strength levels and rate of strength gain of the concrete. While the concrete complexity is increasing, expertise in the DOTs is decreasing due to staff reductions and retirements. The responsibility for developing mixture designs and proportions is being assigned more and more frequently to the contractor by the DOTs. The trend is towards performance-related specifications for paving concrete, because (among other reasons) that approach takes much of the mixture design burden off the DOT, and also gives the contractor added freedom in the mixture design process. As noted, despite the added complexity, the behavior of the cement and other concrete components must be compatible and predictable when combined.

PERFORMACE AND DURABILITY

Although, from a user standpoint, the main performance criteria may be a pavement's ride, there are other aspects of pavement performance that must be considered and will ultimately influence overall serviceability, including ride. The three main categories of performance problems (using jointed plain concrete pavements as an example) are uncontrolled cracking, joint or crack faulting and concrete deterioration[4].

Uncontrolled cracking herein refers to full depth cracking that may be oriented either transverse or longitudinal to the pavement. It usually occurs when tensile stresses in the pavement exceed the tensile strength of the concrete. It is influenced not only by the properties of the concrete (such as strength, shrinkage potential, creep, and elastic modulus) but also by the pavement design (thickness, joint spacing, etc.), construction processes and procedures, and the environment both at the time of construction and over the long term.

Joint (or crack) faulting results from the loss of load transfer across the discontinuity. It is influenced by many of the same factors as uncontrolled cracking, especially those that cause excessive opening of the joint or crack. Faulting will also depend on the amount of slab support provided by base or sub-base layers under the concrete slabs.

Both cracking and faulting are more likely to occur where the potential for volumetric decrease in the concrete after placement and setting is higher. Thus, concretes that are more volumetrically stable during hydration, temperature change and moisture change are desirable for their resistance to cracking and faulting. Proper curing during construction can help reduce the tensile stress build up in the concrete by controlling temperature and moisture gradients and overall temperature and moisture changes in the concrete as it hydrates and gains strength.

The durability of the concrete as a material includes such factors as freeze-thaw distress, alkali-aggregate reactivity, sulfate attack and chemical (de-icer) attack. It is dependent not only upon the concrete components and characteristics, but also on the environment to which the concrete is exposed. Environmental exposure has two important time frames: short-term, that is during the construction process; and long-term, that is over the service life of the pavement. During construction, the concrete needs to be properly cured, that is both temperature and moisture of the concrete adequately controlled in order to assure full development of its durability, particularly in the near surface zone. Over the long-term, the concrete must be properly designed and constructed to resist such deterioration mechanisms as freeze-thaw cycling and sulfate exposure.

The final important aspect of performance is the predictability of concrete performance. With the decrease of expertise in the DOTs, as noted above, it is especially critical that the performance of the concrete is reliable and predictable. In order to instill confidence in their use, the same concretes must behave the same way day in and day out. That is, the actual properties and performance obtained throughout construction must be consistent with what was achieved during trial batching and used as input during the pavement design phase.

COSTS

There are three aspects of concrete cost that are important to the use of concrete in highway construction, particularly in pavements, where large volumes of concrete are used. These aspects are initial cost, life-cycle cost, and competitive cost.

Initial cost is the cost to construct the pavement, and will often dictate how a paving project is addressed, since in a typical pavement management system an agency has a certain number of lane miles that must be attended to every year. The initial cost will therefore impact the approach taken (maintenance, repair, rehabilitation or reconstruction) as well as the material(s) used.

If it is used as a part of the decision-making process, life-cycle cost can often help demonstrate the benefits of concrete due to the usual longer service life, and longer time until first repair or rehabilitation activity. As the name implies, life-cycle cost is the total cost of a pavement divided by the total life of that pavement. It can be calculated in many ways, using a variety of interest rates, so any cost comparisons based on life-cycle costs must be based on the same types of calculations in order to make a fair comparison.

The main competitor for concrete in paving is asphalt. The concrete paving industry is always striving for means to decrease the initial cost of concrete pavements, so they will be more competitive with asphalt pavements on that basis, while not decreasing the advantage which concrete has in service life. This is seen as a means to increase market share. An initial cost decrease can be achieved by using less expensive materials, thinner slabs, or more efficient construction processes.

SUMMARY

The cement and concrete needs of the highway industry are influenced by:

- the currently dominant construction modes of maintenance, repair, rehabilitation and reconstruction;
- user demand for minimum lane closures and rapid construction;
- increased complexity of mixtures and a decrease of DOT experience with and knowledge of the concrete's behavior;
- desired performance and predictability of performance for concrete and concrete pavements;
- initial, life-cycle and competitive costs of cement, concrete and concrete pavement.

HIGHWAY INDUSTRY NEEDS AND HOW WE GET THERE

Based on the above discussion, the highway industry needs for cement and concrete may be highlighted as follows:

- materials that lend themselves to rapid repair, rehabilitation and reconstruction;
- concrete that is easy to mix, place and consolidate, with consistent, predictable set times and property development, including strength gain;
- cement and concrete that are volumetrically stable during hydration, and during environmental changes in temperature and moisture;

- cement and concrete that are more durable as materials, and therefore longer lasting;
- cement and concrete which are lower in cost and which provide pavements which are lower in initial cost and therefore even more competitive with asphalt.

To address a number of these issues, the Federal Highway Administration has developed a cooperative research program with industry which has been designated the Concrete Pavement Technology Program[5,6]. Industry representation in this cooperative effort includes the American Concrete Pavement Association, the Portland Cement Association and National Ready-Mix Concrete Association. The program has as its four main goals:

- Reduce user delays
- Reduce costs
- Improve performance
- Foster innovation.

Operating since 1999, the program is addressing each of these goals through a series of practical, results-oriented research projects. More can be learned about individual studies by visiting the FHWA R&D website at www.tfhrc.gov (look under pavements/concrete pavements) and the industry website at www.IPRF.org. We welcome your comments, suggestions and input on on-going work related to the program activities.

We also look forward to working with industry to explain our needs and cooperate in order to address them.

REFERENCES

[1]1999 Status of the Nation's Highways, Bridges, and Transit: Conditions & Performance, Report to Congress, Executive Summary; June, 2000, U.S. Department of Transportation, Federal Highway Administration and Federal Transit Administration; Washington, D.C.; Publication No. FHWA-PL-00-017; 17 p.

[2]Our Nation's Highways, Selected Facts and Figures; April, 1997; Federal Highway Administration; Washington, D.C.; 24p.

[3]Get In Get Out Stay Out!: Proceedings of the Workshop on Pavement Renewal for Urban Freeways, 2000, Transportation Research Board, National Research Council, National Academy Press; Washington, D.C.; 90 p.

[4]Forster, Stephen W.; 2001; The "Systems" Approach to Concrete Pavements; in 7th International Conference on Concrete Pavements; Proceedings; Volume 2; pp 1013 – 1038; International Society for Concrete Pavements.

[5]Concrete Pavement Technology Program - Developing a New Generation of Concrete Pavements; 2000; Innovative Pavement Research Foundation; Falls Church, VA.; 8p.

[6]Cole, Lawrence W.; 2001; Creating a New Generation of Concrete Pavement; in TR News, Number 215, July-August 2001; Transportation Research Board, Washington, D.C.; p. 15 – 21.

Photo 17: S. Forster

Photo 18: K. Ogden, E. Suderman, B. Suderman and W. Ma

THE IMPORTANCE AND POTENTIAL OF QUANTITATIVE MICROSTRUCTURAL CHARACTERIZATION IN RESEARCH ON CEMENTITIOUS MATERIALS.

Karen L. Scrivener
Laboratory of Construction Materials
Swiss Federal Institute of Technology at Lausanne (EPFL)
1015 Lausanne, Switzerland

ABSTRACT

For many materials the study of microstructure has played an essential role in research aimed at their understanding and improvement. For cementitious materials the great heterogeneity of the microstructure makes the extraction of comparative information difficult. However, because of this, it is important that such characterization should be quantitative rather than qualitative.

Backscattered electron images of polished sections studied in the scanning electron microscope, coupled with local microanalysis, are fruitful techniques for quantitative characterization as illustrated by the example of the study of delayed ettringite formation in mortars cured at elevated temperatures. However, many images must be analyzed, or points analyzed to obtain a statistically valid description of the average microstructure.

Further development of such techniques is possible and it is believed that particularly promising approaches are integration with other techniques, such as quantitative X-ray diffraction (XRD) using Rietveld analysis.

THE IMPORTANCE OF MICROSTRUCTURE.

The microstructure of a material is the key link in understanding the relation between the processes that form the material and its properties. This is the essential tenet of Materials Science. Its application has been very successful in understanding and improving the behavior of a wide range of materials – metallic, polymeric and ceramic. However, despite the use of many microstructural techniques to study cementitious microstructures, this approach does not at present play a central role in research on such materials, much of which remains empirical parametric testing.

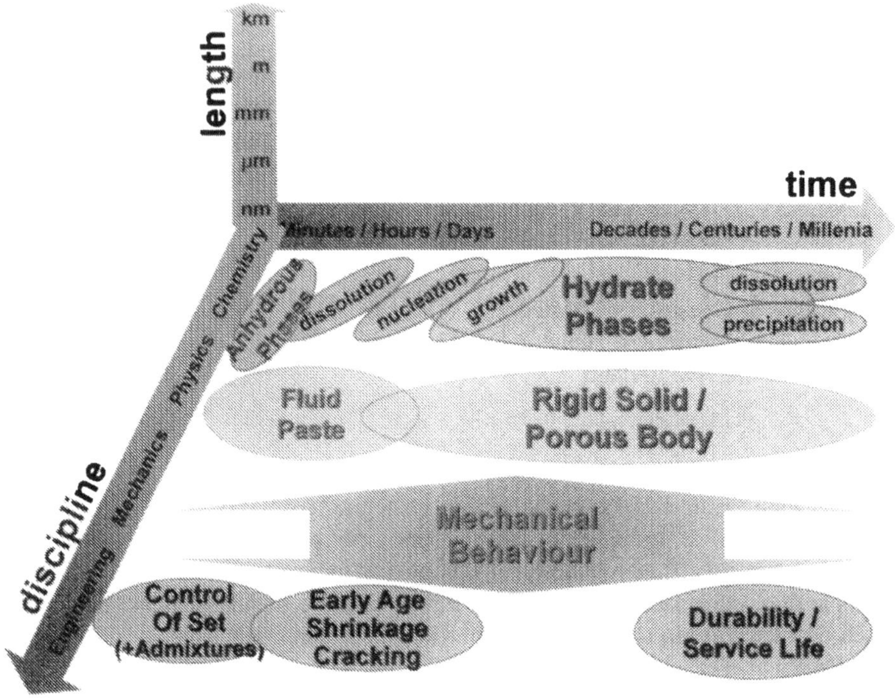

Figure 1: Scheme illustrating the complexity of cementitious materials.\

THE PARTICULARITIES AND PROBLEMS INHERENT IN CEMENTITIOUS MICROSTRUCTURES

Despite being the most widely used material in the world, concrete remains one of the most poorly understood, primarily because of its high level of complexity (Figure 1):

- The range of length scales extends from that of the largest aggregate at several centimetres in size (or even that of the structural element at several metres), down to that of the smallest pores and crystals, which are only nanometres in size – 7 to 10 orders of magnitude.
- On a chemical level more than seven oxide components (CaO, SiO_2, Al_2O_3, Fe_2O_3, SO_3, Na_2O, K_2O, …) must be considered along with water.
- Physically we are concerned with a fluid suspension that transforms into a rigid, porous solid.
- And processes occurring in the first few seconds have an impact on the performance over decades, centuries or even millennia.

Figure 2: Fracture surface of 24hour old cement paste, showing C-S-H and ettringite.

MICROSTRUCTURE = PRETTY PICTURES?

Unfortunately, most microstructural studies stop at the level of pretty pictures, such as that shown in Figure 2 [1].

This is one of my own pictures taken nearly twenty years ago. Such qualitative studies have provided us with much useful information on the development of microstructure during hydration, but due to the very heterogeneous nature of cement paste there is generally as much variability within the same sample as between samples from different cements or prepared under different conditions. Such pictures are lacking in context – they show only a fracture path, a zone of weakness. We cannot see how much unhydrated cement is lying beneath the surface and above all we cannot extract any numbers, which can be used to compare materials produced by different processes, or having different properties.

Despite these limitations, this is still the extent of most microstructural studies reported in publications. As an example, a subjective assessment of the studies by electron microscopy* presented at the recent Euroseminar on the Microscopy of Building Materials[2], indicated that over half remained at this qualitative level. Why is this? The following reasons may be put forward:

* It should be noted that this assessment did not include optical microscopy studies, which were generally of a much less superficial nature, perhaps due to the wider availability of equipment, to the low depth of field of this technique, or to the longer establishment of mineralogical investigations.

- Scanning electron microscopes, although widely available, generally remain shared multi-user facilities, each user has at most only a few hours per month, so it is difficult to do much more than take a few illustrative pictures.
- Study of microstructure is not an integral part of most PhD studies of cement and concrete, but something tacked on at the end.
- The huge leap between the time to take a quick picture of a fracture surface, compared to that needed to perform a more quantitative study is a severe discouragement, when the pressure is to publish as much as possible in as short a time, regardless of what is actually added to our knowledge of the material.

PROGRESS TOWARDS QUANTITATIVE CHARACTERIZATION - THE NEED FOR A STATISTICALLY SOUND APPROACH.

The first requirement for quantitative characterization is to have a representative sample. This is one of the main advantages of optical microscopy, which has a very low depth of field so samples must be polished flat. The study of polished sections in reflected light is widely (and successfully) used for the study of cement clinkers, and the study of thin sections in transmitted light is likewise widely used in concrete diagnosis. However, except in simple cases such as air void analysis, the only way to obtain quantitative data is by tedious point counting. Optical microscopy is also limited by low spatial resolution and lack of local chemical information.

Backscattered electron imaging of polished sections in the SEM provides extremely useful pictures (Figure 3). The reproducible contrast between the different microstructural constituents allows them to be distinguished by image analysis[3,4,5]. For cement pastes, overall measures of the amounts of unhydrated cement, of calcium hydroxide, and of porosity have been obtained[6] by analyzing some 20 images – taking a few hours per sample. Unfortunately, the addition of aggregate to create concrete[†], which is the material actually used in practice, increases the heterogeneity and many more images (100+) must be analyzed to assess the average microstructure[7,8]. Unfortunately, it is difficult to justify the time that needs to be invested in such measurements, unless better automation of the analysis can be developed.

[†] Ironically mortars are generally more heterogeneous and difficult to characterize than concretes. At the levels of aggregate typically used, the surface area of the aggregate is higher for mortars than concretes and this results the cement paste being more heterogeneous.

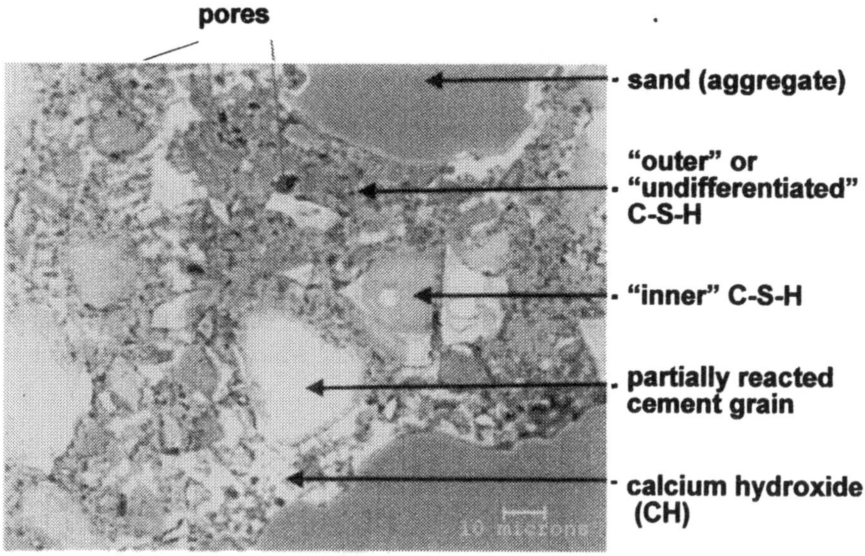

FIGURE 3. Microstructure of mature mortar – polished section, backscattered electron image

In order to extend the usefulness of such techniques, there are other questions to be resolved:

- The resolution of porosity is limited to a few tenths of a micron and the threshold between pores and solid is difficult to define.
- Such images give two dimension sections of a three-dimensional structure, so it is difficult to find meaningful measures of size distribution and impossible to judge connectivity.

Progress in such areas can come from integrating such measures with those from other techniques (e.g. quantitative XRD by Rietveld analysis) and with models which simulate the three dimensional microstructure in a more holistic approach[‡].

[‡] This is one of the major areas of research of the Laboratory of Construction Materials at EPFL (Swiss Federal Institute of Technology at Lausanne), but progress depends on being able to obtain funding for such relatively fundamental work.

COMBINATION WITH LOCAL CHEMICAL INFORMATION – THE MECHANISM OF EXPANSION DUE TO DELAYED ETTRINGITE FORMATION AFTER CURING AT ELEVATED TEMPERATURES

The value of examining polished sections in the SEM can be extended if BSE imaging is combined with local chemical microanalysis by EDS (energy dispersive spectroscopy) of X-rays. Again, it is critical to take a statistical approach as the phases present are often intermixed on a scale finer than the interaction volume of the incident electron beam with the specimen – a volume roughly a micron in dimension within which X-rays are generated. Such an approach is illustrated, by the studies made on the mechanism of expansion due to delayed ettringite formation occurring in mortars which had been subject to elevated temperatures during curing[9,10].

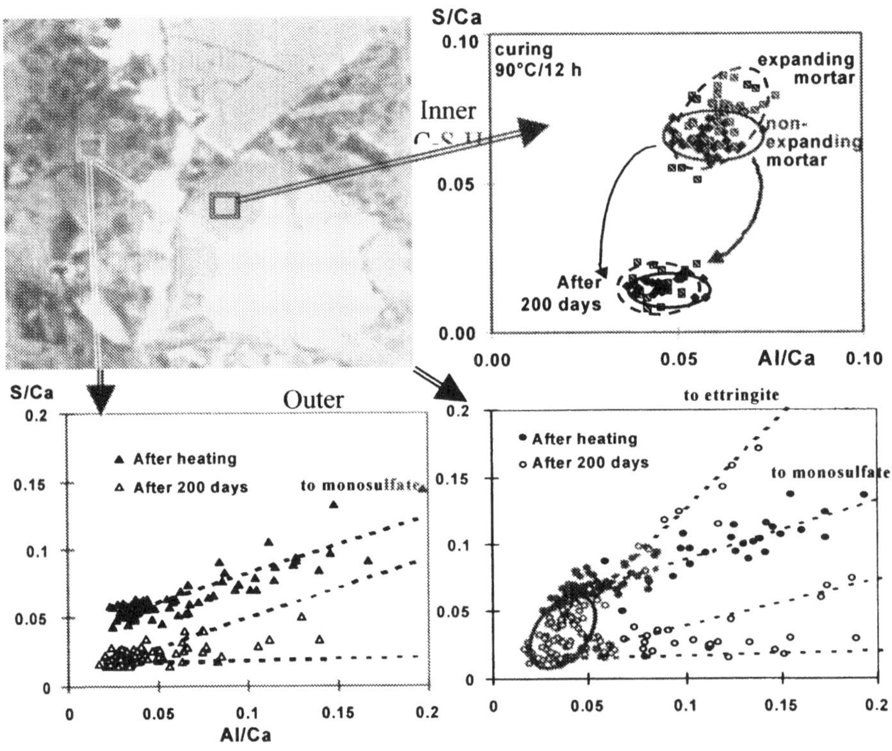

Figure 4: Microanalyses of mortars subject to elevated temperatures during curing, indicating that expansion occurs when ettringite forms in small pores within the outer C-S-H product.

Figure 4 summarizes this study. On each sample, more than 50 analyses were made in carefully chosen similar locations, but distributed over the whole area of the section – namely in "inner" C-S-H product, the hydration shells surrounding partially reacted cement grains; or in "outer" product in the areas between the relics of the larger cements grains, but avoiding deposits of calcium hydroxide.

Due to the large number of analyses taken in each area in each sample, it is possible to deduce which phases are intermixed and the compositions of the individual phases through graphical representations, such as the S/Ca ratio against the Al/Ca. The analyses for the inner product (top right) are similar for both expanding and non-expanding mortars. However, in the analyses of the outer product for the expanding mortar a particular population of points (circled) corresponds to fine mixtures of ettringite within the outer C-S-H. This finding is consistent with thermodynamic arguments [11], which indicate that the pressure that can be generated by crystallization is inversely related to the size of pores in which the crystals form.

SUMMARY

- In this short paper the aim has been to show how, despite their complexity, the microstructures of cementitious materials can be characterized quantitatively and so provide important insights into the relation between processing and properties.
- The rigorous application of such methods, in particular integrating results from different techniques and automation have considerable further potential for development.
- Such approaches are believed to be essential for exploring the potential of special cementitious materials for demanding applications.

REFERENCES

[1] Scrivener, K.L. " The Development of Microstructure during the Hydration of Portland Cement", PhD Thesis, University of London 1984

[2] 8th EuroSeminar on the Microscopy of Building Materials, Athens, September 2001

[3] Scrivener, K.L. and Pratt, P.L., "Backscattered electron images of polished cement sections in the scanning electron microscope", 145-155. (Proc. 6th Int. Conf. Cement Microscopy, Albuquerque, 1984)

[4] Scrivener, K.L. and Pratt, P.L., "The characterisation and quantification of cement and concrete microstructures", 61-68. (Proc. RILEM 1st Int. Cong., Versailles, 1987)

[5] Scrivener, K.L., "The Microstructure of Concrete" in Materials Science of Concrete I, Skalny, J.P. ed (American Ceramic Society, Columbus, Ohio) 1989, 127-161

[6] Scrivener, K.L., Patel, H.H., Pratt, P.L. and Parrott, L.J., "Analysis of phases in cement paste using backscattered electron images, methanol adsorption and thermogravimetric analysis" in Microstructural Development During the Hydration of Cement, 67-76. (Proc. Mat. Res. Soc. Symp., 85, 1987)

[7] Scrivener, K.L., Crumbie, A.K. and Pratt, P.L., "A study of the interfacial region between cement paste and aggregate in concrete" in Bonding in Cementitious Composites, 87-88. (Proc. Mat. Res. Soc. Symp., 114, 1988)

[8] Scrivener, K.L. and Pratt, P.L., "Characterisation of Interfacial Microstructures" pp3-17 in Interfacial Transition Zone in Concrete, ed J.C. Maso, State of the Art Report, RILEM Committee 108 ICC (F.N. Spon,) 1996,

[9] Famy, C., Scrivener, K.L., Brough, A.R., Atkinson, A., Lachowski, E., "Characterisation of C-S-H products in expansive and non-expansive heat-cured mortars: an electron microscopy study". Proceeding of the 5th CANMET/ACI International Conference of Concrete Durability, Barcelona, Spain, Vol. 1, pp. 385-402.

[10] Famy, C. and Scrivener, K.L., Atkinson, A. and Brough, A.R. "“Influence of the storage conditions on the dimensional changes of heat-cured mortars", Cem. Concr. Res **32**(2) 269-278 (2002)

[11] Scherer, G.W. Crystallisation in pores, Cem. Concr. Res. 29(8) 137-1358 (1999).

Photo 19: K. Scrivener

Photo 20: A. Olek and J. Olek

Photo 21: J. Kirkpatrick

Photo 22: J. Pierce and M. Bergin

INTERDISCIPLINARY RESEARCH, AUTHORITY INVOLVEMENT AND TARGETED TECHNOLOGY TRANSFER: THE EXPERIENCE FROM THE DEVELOPMENT OF SELF-COMPACTING CONCRETE IN SWEDEN

Åke Skarendahl
Swedish Cement and Concrete Research Institute
100 44 Stockholm
Sweden

ABSTRACT

Connecting research and market implementation in an innovation process often requires a broad range of input from many actors. Interdisciplinary research efforts have to be made, authorities, clients, production equipment industry, designers etc. have to be involved. The time frame is often a key factor. The implementation of self-compacting concrete on the Swedish market is taken as a case history on conditions and questions that have to be taken into account in the launching of an innovation that will lead to a shift in a mainstream technology.

INTRODUCTION

When introducing new research findings into an innovation process, input from various research disciplines have to be considered and the scope in general has to be widened. The need to shorten the lead-time from initial mechanism research to routine applications will become evident, leading to the necessity of an early and strong involvement of all actors in the process. Especially important is the involvement of clients and authorities, and in most cases, also the end consumers.

The innovation process comprises both the generation of new knowledge as well as the transformation of new knowledge into products, processes and services offered in the market. Every innovation process will have its characteristics, but common features exist. Below, the technology development as well as the market development of self-compacting concrete in Sweden is discussed as a reference case history. Further reading giving information on the development of the self-compacting concrete technology can be found in the references.[1,2,3,4]

CHALLENGES IN INNOVATION PROCESSES

Most innovation processes are based on development of basic knowledge within a narrow and focused research area. If the findings are evaluated to be interesting enough to form a basis for an innovation process, research from supporting disciplines have to be added and applied research, as well as development work, has to be carried out. Identifying the need of new knowledge, mapping existing relevant knowledge, and creating a networking interdisciplinary research activity is a challenge in any innovation process.

Innovations are sometimes driven by technology development as such, but more often the market pull mechanism is the dominating force. Understanding the requirements of the customers, the conditions in the market, existing alternatives, etc is however necessary, whatever the development mechanism. Input is needed from research areas quite different from areas where basic technology mechanism research is carried out. Research, for example, in behavior science, medicine, economics and other non-technical fields will often be necessary. Furthermore, involvement from end-users and investors, as well as from authorities, is often crucial. Identifying the actors that need to be involved in the process, and finding ways to include their involvement, is the second challenge in every innovation process.

Still another important factor in innovation is time. There is a tendency that the accepted time of return on investment in research and innovation is getting shorter and shorter. Furthermore, the development has to be, and stay ahead of, competing developments. Information transfer is gradually getting easier and faster through information technology developments. All these conditions require the components of an innovation process to be carried out concurrently in order to shorten the lead-time from initial idea generation to the launching of the new service, product or process into the market. The time factor is thus a third important challenge in innovation processes.

THE CASE OF SELF-COMPACTING CONCRETE

The development of self-compacting concrete for the Swedish market can be looked upon from an innovation process point of view. The process is characterized by interdisciplinary work, the involvement of many actors, and a concurrent approach.

An increased understanding of particle packing, rheology of suspensions, surface forces and water retention opened up the possibility of significantly improving the workability of fresh concrete. Together with this increasing knowledge, new improved products used as constituent materials in concrete were offered to the market making implementation possible. The emerging understanding of the mechanisms determining flow of fresh concrete coincided with an increasing interest in the market for the need to facilitate the concrete construction process. The possibility of eliminating compaction when casting concrete became the target, which offered interesting visions from various perspectives.

OBJECTIVE OF THE DEVELOPMENT

The initial research and development work on self-compacting concrete was done in Japan in the late 1980s based on what can be described as a market pull mechanism. An analysis of quality assurance aspects in concrete construction pointed to the fact that some advanced civil engineering structures were very complicated to cast due to limited access, congested reinforcement, etc. This created a risk of premature durability failures, reductions in load bearing capacity and aesthetics due to inadequate compaction. An additional concern was an insight that recruitment of skilled workers was becoming increasingly difficult, adding to the quality assurance problem. The conclusion was that development should be driven toward concrete that did not need any extra external compaction, i.e. a self-compacting concrete.

In the Swedish development of SCC, the work did not focus primarily on avoiding the risk of inadequate compaction of advanced and complicated structures but rather on the possibilities for increased productivity and improved the working environment within the bulk volume of concrete applications. The development was initially based on a technology push mechanism where the research findings were pointing at a market development opportunity.

Figure 1. Casting of self-compacting concrete in a regular house-building project.

Productivity

In the constant drive to improve the economy in concrete construction, improved productivity is essential. This is vital along the whole spectrum from low-grade applications like backfill etc. up to the most advanced and technically complex structures. The possibility of complete elimination of the compacting work increases the speed and reduces the cost of construction. In order to exploit the potential of increased productivity, contractors are strongly involved in the innovation process. As the number of concrete workers is reduced and the character of their input modified, the trade unions also play a role in the process.

The increase in productivity might be even more pronounced if the borderline between supplier and contractor is modified. With SCC it will often be applicable to have the concrete supplier not only delivering the concrete to the site but also cast the concrete. The ready mixed concrete company is thus delivering a cast structure rather than a fresh concrete. The legal aspects related to such borderline movements have been shown to be negotiable by the actors.

Figure 2. The borderline between companies is moved as in this case when the pump operator is casting the concrete structure rather than simply delivering fresh concrete.

Self-compacting concrete can also result in improved productivity in the process of transport and placing of the fresh concrete. Utilizing the flow properties of the liquid state, which self-compacting concrete can offer, allows development of new pumping and formwork filling procedures. Thus, participation in the innovation process by pump manufacturers and formwork system designers is vital.

Working Environment

The working environment impact is high in concrete construction. One of the most severe problems is the vibration from handheld pokers causing a blood circulation disturbance known as "white fingers". Another severe issue is the noise caused by the vibration process, either from handheld vibrators, surface vibrators or form vibrators. In addition, the physical demand from moving pump hoses, handling vibrator equipment, etc., is also significant. Since improvements in the working environment is a vital part of the objectives of SCC, researchers in medicine and specialists in working environments are involved in the innovation process.

Improvement of the working conditions is essential as a means of increasing total cost efficiency of concrete construction. In many regions, the average age of concrete workers is high, requiring specific attention to working conditions. The development of self-compacting concrete is likely to increase the attractiveness of concrete construction, thus increasing the possibilities of recruiting qualified staff. It is important to erase the conception of concrete work being dirty, dangerous and difficult.

Elimination of the use of vibrators reduces the noise on building sites, which is an advantage for both the working environment as well as the disturbance for building site neighbors. Expertise in the measurement of noise is a vital component in the innovation process.

MATERIAL MECHANISMS

Fresh concrete can be described as a particle suspension, although rather complex since it has an extraordinarily large range in particle sizes and the liquid phase is continuously changing due to hydration. As in all particle suspensions, obtaining flowability without segregation is the key goal. Self-compacting concrete is extending the flowability, while maintaining stability of the suspension, to such a degree that the concrete can flow through openings, fill formwork and consolidate itself without any added external energy input. To achieve this behavior – while considering casting of reinforced components - the following three requirements related to the workability of the material have to be met:

- filling ability
- resistance to segregation
- passing ability

The main mechanism required to achieve appropriate filling ability is a reduction in inter-particle friction. The most important means are reduction of surface tension (obtained e.g. with super plasticizers), and optimum packing of fine particles (normally obtained through the use of fillers). Behavior is evaluated through the study of rheological properties.

Following the Bingham model (Figure 3), a low yield stress value is normally aimed at for SCC while the target plastic viscosity will vary with production technique and application. The thixotropy of the suspension is also an important property as it influences planning of the casting sequence, as well as formwork design. Paste, mortar and concrete viscometers are used to evaluate the rheological properties.

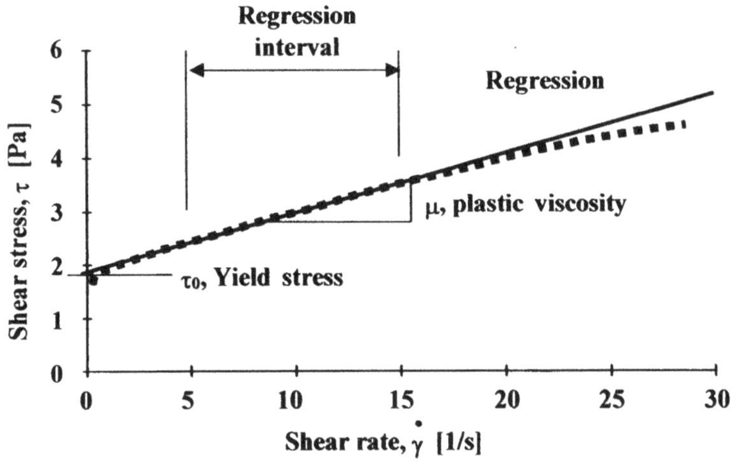

Figure 3. Rheological evaluation following the Bingham model of the relationship between shear stress and shear rate gives yield stress (at rate = 0) and plastic viscosity (inclination of curve).

The rheological behavior is, of course, also central when it comes to the creation of a suspension with sufficient resistance to segregation. The amount of moveable water has to be limited to minimize bleeding, and the liquid phase has to be so composed as to give the suspension a certain plastic viscosity behavior.

The functional requirement related to passing ability, or in other words the avoidance of aggregate blocking, will depend on the reinforcement arrangement for the specific structure to be cast. The size and shape of the coarse aggregate, as well as the volume of the liquid phase, are other important parameters related to blocking.

Knowledge from several research areas is involved in the studies of these material mechanisms. The core research is connected to the rheology of particle suspensions. Surface chemistry, mineralogy and particle packing modeling are important interdisciplinary contacts. The chemical and mineral industries are playing an important role in the innovation process.

MIXING, TRANSPORT AND PUMPING

There seems to be a good possibility of producing self-compacting concrete in most types of mixers, even if some mixer types have better potential than others to produce a good mix. The order of adding the constituents to the mixer is vital to achieve good conditions for dispersion of the high amount of fine particles normally used. To be able to fully optimize SCC, research on powder dispersion is essential, as is participation of the mixer producers.

The quality of the constituents have to be kept under strict control, in particular the moisture content of the fine aggregate since the flowability is sensitive to the total water content in the mix. Knowledge on moisture measurement and control is thus necessary, as well as the supply of suitable equipment. The importance of suitable filler materials implies a need for knowledge on mineralogy and mineral processing, as well as investments in silo capacity at batching plants. Normal transport procedures can be used as well as traditional concrete pumps. Control of the pumping process, however, is slightly different, as the rheological properties of most fresh SCC will cause a different response in pump pressure versus feeding rate. Research based on fluid mechanics is essential to be able to control pipe transport, ease of flow, etc.

CASTING

Placing of self-compacting concrete can be done with the same type of equipment as used for normal concrete. However, avoiding all need for compaction work allows for alternative methods. Pumping the concrete directly into formwork through valves is the common practice, as is casting in enclosed spaces. Sensor systems for the control of the flow are used.

One of the most significant areas of development of SCC-technology is planning and control of the entire casting process. Utilizing the material properties of SCC to modify the casting process has led to substantial improvements in productivity on many projects. The possibility of using distinct element modeling (DEM) to theoretically simulate flow is under development. Here, expertise from the geomechanical area is of great value.

Self-compacting concrete is flowing and moldable while in motion. In order to get a fresh concrete still in motion integrated with a previous pour now at rest, the time between layers of concrete can may not be too long. This requires a balance between the thixotropy of the concrete and the casting process. The performance of the cements and their interaction with the admixtures used are important in this respect.

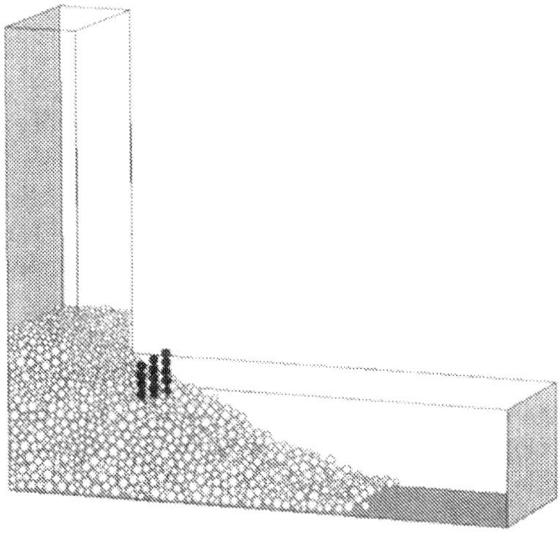

Figure 4. Modelling of flow of fresh SCC through DEM technique.

Thixotropic behavior is also important in reducing form pressure and thus allowing the use of more slender and cheaper formwork

Another aspect of importance is the tightness requirements on the formwork. If sufficient water retention properties are built into the material, casting with SCC is less sensitive to water seepage through formwork openings compared to what is the case in an ordinary casting.

HARDENED CONCRETE PROPERTIES

The microstructure of self-compacting concrete is denser than that of a corresponding normal vibrated concrete for two main reasons. First, the mineral surface area is enlarged from the use of an increased fine particle content and, second, the interfacial transition zones are less pronounced following the avoidance of vibration. These effects have been verified by studies on the microstructure.

The improved microstructure is likely to improve mechanical as well as durability properties. This is confirmed from the practical use where increased strength as well as reduced transport through the pore system is recorded compared to a vibrated concrete with the same water/cement ratio.

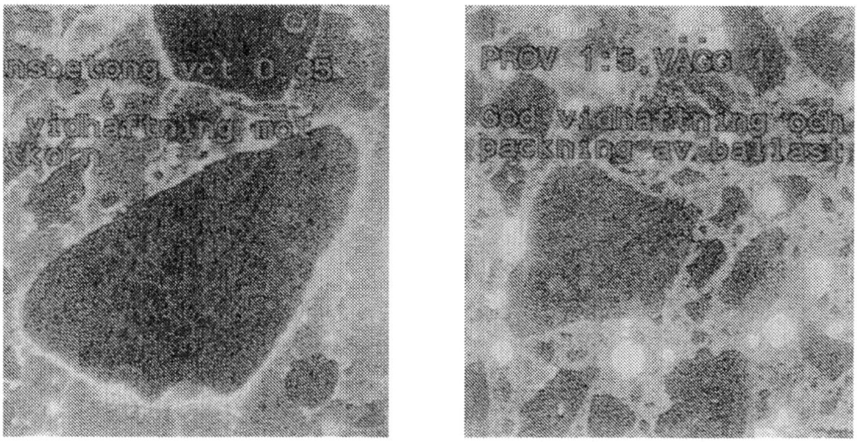

Figure 5: Thin sections of vibrated concrete (left) and SCC (right) indicating a less porous transition zone around the aggregates for SCC than for vibrated concrete.

Experience gained from projects on high as well as on moderate strength concrete shows that the surfaces obtained have less porosity and are more even in color than what is common for normal concrete. The smoothness of the surface was especially evident in tests where the concrete was pumped from the bottom of the formwork through valves.

The improved properties in terms of strength, durability and surface quality are adding to the user values of SCC in terms of reduced binder content or more slender structures, longer service life and improved aesthetics. Consequently, it contributes to an improved sustainability of concrete construction.

CODES AND STANDARDS

The introduction of new materials and new processes into the market has to take existing systems for standards and codes into consideration. Innovation processes therefore also have to involve a strategy on how an innovation shall relate to these systems. As many of the components of codes and standards in construction are still prescriptive rather than based on functional requirements, implementation of new technology is often difficult.

The implementation of SCC in the Swedish market was first made in the area of civil engineering structures, i.e. structures under strict regulation in a number of aspects. Close co-operation with one of the leading authorities in construction, the Swedish National Road Administration, opened up the possibilities of having a continuos and open dialog on the possible influences of the new technology on the code and standard system. This made it possible to obtain provisional acceptance for full-scale applications. With increasing experience from full-scale work, confidence in the technology widened and the acceptance of authorities and clients in general grew.

Following a committee work involving authorities, clients, contractors, designers and the material industry, modifications to existing codes and standards for SCC use will be published in late 2001. This will further strengthen market acceptance.

MARKET ACCEPTANCE AND APPLICATION AREAS

The use of self-compacting concrete in on-site construction in Sweden was approximately 3 % of the market in 2000. This figure is likely to at least double during 2001. The development in precasting is also strong and in the near future some factories are likely to have completely changed to SCC. The interest has been very strong for inclusion in civil engineering applications but the use in normal house construction is now increasing rapidly.

Self-compacting concrete is a more difficult and complex technology, when compared with normal concrete. It requires improved skill and competence in all parts of the process. Education and training thus have to be given a very high priority in the development of the practical application of the technology. The cost savings and other positive features obtained, however, will easily outweigh the increased efforts that must be made to cope with the technology. This tendency is very clearly seen in the Swedish market. There are very strong reasons to believe that SCC will be the mainstream concrete technology within the near future.

REFERENCES

[1]JSCE. Proceedings of the International Workshop on Self-Compacting Concrete. Kochi. Ed. by Ozawa K and Ouchi M, JSCE Concrete Engineering Series 30, 1998, 400 pages.

[2]RILEM TC 174-SCC. Self-Compacting Concrete. Proceedings of the First International Symposium . Ed. by Å Skarendahl and Ö Petersson, Stockholm, September 1999, 790 pages.

[3]RILEM, Self Compacting Concrete. State-of-the Art Report of RILEM TC 174-SCC. RILEM Publications SARL, Paris, 2000, 152 pages.

[4]Proceedings of the Second International symposium on Self-Compacting Concrete. Ed by K Ozawa and M Ouchi. Tokyo October 2001. COMS Engineering Corporation, Kochi, 2001, 740 pages.

Photo 23: Å. Skarendahl

Photo 24: R.D. Hooton

Photo 25: J. Lukasik, C. Andrade and J. Olek

Photo 26: K. Scrivener, J. Marchand, T. Coverdale and S. Diamond

A NEW MODEL OF ORDINARY PORTLAND CEMENT HYDRATION DERIVED BY MEANS OF ESEM-FEG

B. Möser and J. Stark
Bauhaus Universität Weimar
F. A. Finger Institut für Baustoffkunde
Coudraystraße 11
D-99421 Weimar, Germany

ABSTRACT

The visualization of structures resulting from processes taking place within a few nanometers is immensely important when studying the properties of concrete and mortar. Electron microscopy is in principle capable of obtaining images of structural constituents in this size range. Unfortunately, studies with conventional electron microscopes are subject to serious limitations: the specimens must be high-vacuum-stable or correspondingly prepared and the specimen must generally be electrically conductive or be made conductive by a coating of the surface. For these reasons, an Environmental Scanning Electron Microscope with Field Emission Gun (ESEM-FEG) in different operating modes has been used to investigate wet materials in their native state. These methods provide deep insights into the moist micro and nano world. It even makes it possible to carry out sophisticated observations of hydration processes in complicated multi-phase systems, such as Portland cement. The article describes the investigations of the phase development during the hydration of OPC from the induction to the final period. The results are summarized in a scheme for a new model of ordinary Portland cement hydration.

INTRODUCTION

The model of cement hydration developed by Powers and his colleagues (1953) is now considered as the first pioneering draft scheme. The hydration products of Portland cement were initially described as spherical [1]. With the first electron microscope micrographs, the concept of spheres was replaced by the model of groups of randomly orientated and layered units (cement gel). The cement gel was described as a solid substance with an average pore diameter of approximately 1.8 nm [2]. On this basis, Powers developed a model for the creep and shrinkage of hardened cement paste. Feldmann and Sereda (1968) investigated the different types of water bonding to hydration products. C-S-H phases - the main hydration products - were described as corrugated platelets [3].

Kondo and Daimon (1974) presented a model that conformed the two above-mentioned models [4]. As early as 1965, Richartz and Locher [5] introduced two different types of C-S-H phases (type I and II). In the C-S-H (I) phase, the C/S ratio was said to be approximately 0.8 to 1.5 and in the C-S-H (II) phase, it should be between 1.0 and 2.0. According to this model, the crystallinity of the phases decreases with increasing C/S ratio. C-S-H (I) is shown in electron micrographs as flakes, while C-S-H (II) forms fibers. In this model the fibers corresponded to rolled platelets with layered inclusions of CH (Fig. 1). According to Locher, Richartz and Sprung, as well as Sylla and Rechenberg [6-9], the observed C-S-H phases grow initially as long fibers and later as short fibers.

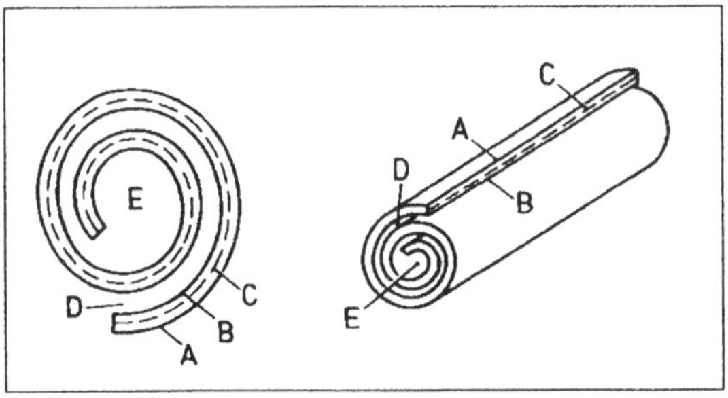

Figure 1: Rolled C-S-H (II) platelets (A and B: C-S-H layers; C: interlayer region; D: space between the rolled leaves; E: internal space in the roll according to Richartz and Locher [5].

To clarify the interaction of the hardened cement paste with water, Setzer (1977) described by means of a model the maxima in the pore radius distribution of hardened cement paste to be around 2.5 and 7.0 nm [10]. In the same year, Wittmann introduced the hydration products of Portland cement as xerogels in which each particle is linked to its surroundings by different bonding forces [11]. Taylor (1992) designated the crystal structure of the hydration products, the C-S-H phases, as “similar to tobermorite”[12]. According to sheet silicates, tobermorite consists of calcium layers located between silicate chains. Taylor also compares C-S-H phases with the mineral jennite, which has a higher C/S than tobermorite and is said to consist of distorted layers of CaO.

As a result of a very intense evaluation of the literature, Massazza and Daimon (1992) pointed out five different morphologies of C-S-H phases that grow during the hydration process [13]. There are descriptions of gel-like, needle-like and sword-shaped C-S-H phases as well as of flake-shaped phases. The morphological changes of the C-S-H phases during the hydration of C_3S were shown for the first time in detail by Meredith et al. (1995) based on investigation with an Environmental Scanning Electron Microscope (ESEM) [14].

Jennings (2000) proposed a new model for the microstructure of calcium silicate hydrates in cement paste [15]. He referred to the "tobermorite-jennite model" by Taylor and proposed two types of C-S-H: LD C-S-H (low-density) which crystallize in the large capillary pore space during the main hydration period, and HD C-S-H (high density) which is formed in the significantly more restricted voids around the C_3S particles.

It can be concluded that the existing theories on the morphology and structure of C-S-H phases arise from different kinds of investigation methods (mainly electron microscopy and NMR studies). Often, these results are difficult to compare because of the lack of a unified classification of the hydration products and because of the artifacts produced during sample preparation for electron microscopy.

MATERIALS AND METHODS

Fortunately, the preparation of specimens to be studied with the ESEM is uncomplicated, unlike that for conventional scanning electron microscopy, where drying and coating of the specimens is essential. In most cases with the ESEM it is unnecessary to stop the hydration process. The educts used for our investigations are synthetic C_3S, ß-C_2S, and ordinary Portland cement (OPC). All pastes were prepared with a water-cement ratio of 0.3. This value has proven to be favorable for most types of cement. Higher water-cement ratios result in sedimenting out and bleeding structures and thus initiate a thinning effect: readily soluble salts - like K_2SO_4, Na_2SO_4 - are dissolved in the excess fluid and settle on top of the specimen, i.e. they are not accessible for the hydration reactions.

The hydration of the cement and the synthetic clinker minerals were performed outside the microscope in airtight sealed containers, at room temperature. While the paste was still plastic, a roughly pea-sized specimen was taken at preset times and applied to the sample holder on a Peltier cooling stage of the ESEM. After setting the parameters inside the microscope chamber to the adequate pressure and temperature (to reach the desired humidity of 80-95%), the specimen was fractured with the aid of a micromanipulator. This fracturing is very important because, during the microscope adjustment to the relative humidity of 80-95%, a skin consisting of alkali salts and presumably also carbonates is formed on the specimen surface. After the first setting of the paste, fracture surfaces produced outside the microscope have been used for the examination.

The ESEM-FEG has been used in different operating modes: wet-[ESEM], low- as well as high-vacuum and low- and high-voltage modes. To identify the chemical composition of singular hydration phases we applied Energy Dispersive X-Ray Spectrometry (EDX) combined with the ESEM.

To monitor the hydration progress continuously time resolved, we measured the evolved reaction heat with Differential Calorimetric Analysis (DCA). Powder-X-Ray-Diffraction (XRD) was applied to confirm the EDX-Analysis of the minerals.

RESULTS

Hydration of C_3S (alite)

Calcium silicate hydrates (C-S-H phases) of variable chemical composition, as well as portlandite (CH), are formed during the hydration of C_3S. The hydration reaction can be divided into 5 time periods on the basis of the DCA (Fig. 2). According to these five periods, changes in the microstructure have been investigated with the ESEM. Systematic ESEM observations show typical morphological changes for each hydration period. These have been summarized in Table 1.

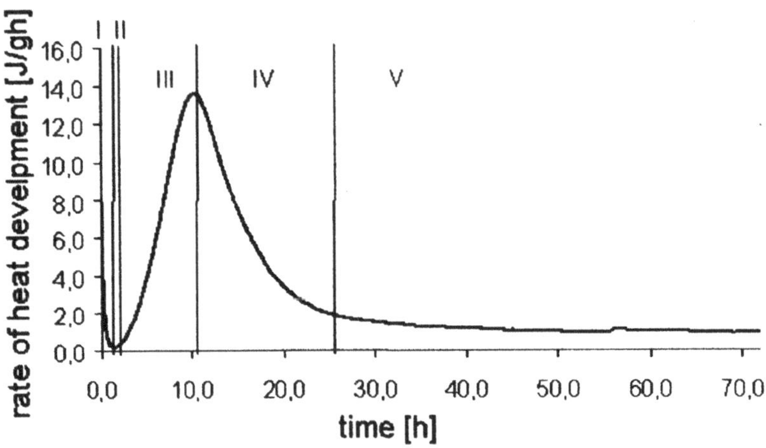

Figure 2: Heat rate shown in hydration periods (Table 1) of C_3S (DCA measurement).

It can be assumed that at the beginning of the acceleration period all C-S-H phases grow in the form of fibers. These fibers grow in length during the hydration process (Fig. 3) with hardly any change in thickness. In compacted concrete, the C-S-H fibers appear to grow together. They also interlock. This minimizes the pore space between the fibers. In fact, the pore diameter of these fibers must be in the range of the size of gel pores.

	PERIOD	TIME	REACTION KINETICS	CHEMICAL PROCESSES	PHASE DESCRIPTION
1	Induction period / Initial hydrolysis	first minutes after mixing	chemically controlled, rapid reaction	start of hydrolysis, ions go into solution	formation of a smooth reaction layer around the grains
2	Dormant period	approx. 20 min to 2h	core and diffusion controlled, slow reaction	very small chemical activity	first heterogeneous nucleation
3	Acceleration period	2h to 11 h	chemically controlled, rapid reaction	start of the formation of hydration products	µm-large, thin laminated portlandite crystals as well as short needle-like C-S-H fibers
4	Deceleration period	11 h to 26 h	chemically and diffusion controlled reaction	continuous formation of hydration products	one-dimensional growth of the C-S-H fibers, aggregation of this fibers, stronger interlocking of the particles; intergrowth of the platy CH crystals
5	Final period	after approx. 26 h	diffusion controlled reaction	slow formation of hydration products	further growth in length of the C-S-H fibers, interlocking

Table 1: Stages of C_3S hydration (source: Mindess and Young [16])

Changes in reaction conditions have an extreme effect on the C_3S hydration: A drop in temperature from 22°C to 5 °C, for example, results in a retardation of the hydration reaction by several hours.

Conventional Scanning Electron Microscope (SEM) imaging requires samples that are stable in high vacuum and are electrically conductive. The preparation steps necessary (drying, coating or sputtering, etc) prevent noninvasive observation of the C-S-H structures (Figs. 4 and 5).

Figure 3: Longitudinal growth of C-S-H fibers during the C_3S hydration process.

Figure 4: Formation of needle-like C-S-H phases during the hydration of C_3S. After 600 days hydration time the fibers grow up to a length of 1.5 µm and the needle tips have a diameter of only approx. 5 nm.

Figure 5: The same sample as in Fig. 4, but with an approximately 30 nm thick layer of carbon for imaging in conventional SEM. The morphologies of the C-S-H phases are changed drastically by thermal stressing during the coating process and by a vapor-deposited layer.

The morphologies of the C-S-H phases are changed drastically by the thermal stress during the coating process and by the vapor-deposited layer itself. Furthermore, we doubt that the chemical compositions given for C-S-H (I) and C-S-H (II) determined with EDX can be very precise. The C-S-H phases are nanometer scaled needle-shaped crystals (crystal thickness of a few nanometers). The excitation volume of electrons with an accelerating voltage of 15 kV is about a few cubic micrometers. Monte Carlo simulation calculations for a beam voltage of, for example, 15 kV give an excitation depth of approximately 2.5 μm, if the density of C-S-H phases - about 2.2 g/cm^3 - is applied. For comparison, the low energy operation of 2.5 keV yields a high spatial resolution for the X-ray microanalysis between 50 – 100 nm, which is related to the approximate thickness of a single C-S-H phase (see Fig. 4 – 8).

Thus, we have to admit that the former measurements made with a conventional SEM / EDX give only a rough estimation of the C/S ratio of the small C-S-H phases. Analysis of uncoated and non-conducting specimens on the nanometer scale can only be obtained by working in the range of low accelerating voltage (low voltage SEM with field emitters) in combination with a sensitive, high resolution X-ray spectrometer. The use of a standard SEM equipped with a tungsten filament is not advisable for producing high resolution images at these voltages and magnification levels. Further prospective methods are ultra high resolution EDX analysis or Electron Energy Loss Spectroscopy (EELS) in the Transmission Electron Microscope (TEM).

Although many classical investigations on the formation and composition of the cement hydrates have been correct in their time with the available methods, now they cannot stand a critical examination. It is now clear and widely accepted that C-S-H phases are nanoscale fibers but not rolled platelets or flakes. The frequently used designation C-S-H "gel" is quite correct from a physical point of view as a gel implies a particle size. From the point of view of cement chemistry and mineralogy, however, the C-S-H phases involved are definitely not gels (term derived from colloid chemistry) but are more probably highly disordered nano-crystalline minerals due to their large surface areas. This assumption is further supported by XRD measurements, where only very weak interferences occur. The term "amorphous to X-rays" is a very appropriate expression in the case of C-S-H phases.

The formation of the C-S-H phases as long fibers initially and later as short fibers is also not sustainable [6]. In fact, the C-S-H phases tend to start as short fibers and grow to a length of approximately 1.5 μm as hydration progresses (Fig. 3). The ESEM images in Figs. 6 and 7 show the formation of the microstructure at selected times during the hydration process at room temperature. Over a time period of 21 days, the C-S-H phases grow continuously to a length of 900 nm, the needle tips have a diameter of only approximately 5 nm. The longitudinal growth starts from the surface of the clinker grain. As a result of this merely one-dimensional growth, C-S-

H phases develop a tight interlocking due to their sharply tipped, needle-like shapes (Fig. 7). In very dense structures, we assume an intergrowth between the fibers as well as interlocking of the fibers with one another.

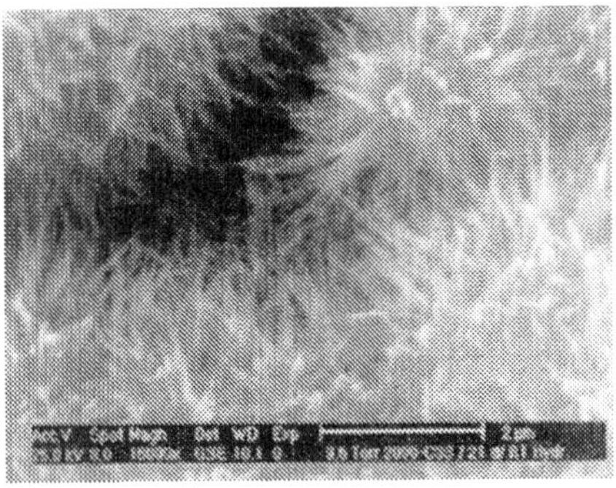

Figure 6: C_3S hydrated for 21 days: tipped, needle-like C-S-H phases (length up to 900 nm).

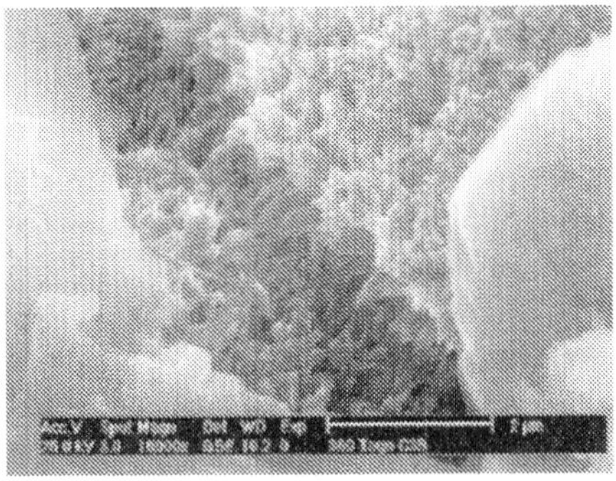

Figure 7: C_3S hydrated for 360 days: interlocking of the C-S-H phases (length up to 1200 nm), portlandite right and left.

Hydration of ß-c_2s (Belite)

The same fiber-shaped hydration products as formed during the C_3S hydration are observed for the hydration of ß-C_2S. However, as already often reported by others, this hydration occurs significantly slower than that of the C_3S. After a hydration time of about one year the fiber size in hydrated C_3S and ß-C_2S are equal. Extremely long C-S-H fibers (up to 2.5 µm) were imaged in an investigation of ß-C_2S which hydrated for 3 years (Fig. 8). This length growth is expected to be a contribution to the secondary hardening.

Hydration of Ordinary Portland Cement (OPC)

Investigations of the phase development during the hydration of OPC from the induction period to the final period (Table 1) are summarized in Figure 19. This scheme for the normal hardening process of OPC at room temperature was derived from investigations of Stark, Möser, Eckart [17] and Hader [18]. Here we would like to give only a short overview of the hydration behavior of OPC with a low (ESEM-images on the left side) and a high (ESEM-images on the right side) content of water soluble potassium.

Figure 8: ß-C_2S hydrated for 3 years: formation of extremely long (up to 2500 nm) C-S-H fibers – here in front of a portlandite crystal . Note: the strength contribution made by the huge portlandite crystal is likely to be small or portlandite can even act as a slip plane under load.

The hydration of industrial ordinary Portland cement differs significantly from the hydration of individual clinker phases. A different chemical equilibrium is set up in the aqueous phase as a result of the interaction between the hydration reactions of several adjacent phases, and this has a significant effect on the formation and sometimes on the morphology of the hydration products.

The behavior with time of OPC hydration can, like C_3S hydration, be divided into 5 time periods:

Induction period: Immediately after mixing with water, very small ettringite crystals were formed. Ettringite is the first product formed immediately after mixing. The small prismatic ettringite crystals have a length between 300 and 500 nm and a thickness between 50 and 250 nm (see Figs. 9a and b). If a high quantity of potassium sulfate (mainly arcanite) is available, calcium-potassium sulfate double salts, like syngenite [K_2SO_4 $CaSO_4$ H_2O], crystallize. Syngenite forms long (up to 10 μm) lath-shaped crystals and can stiffen the cement paste (Fig. 9b).

Dormant period (low chemical activity): Syngenite formation was also observed in cements with a very low potassium content but at a later hydration stage (after about 30 minutes). Figure 10a shows the microstructure of this low potassium cement after 2 h hydration time: bent syngenite crystal aggregates and ettringite with the same morphology as after 10 minutes are observable.

Figure 9a: Microstructure after 10 min hydration time [low potassium content]: Only small hexagonal ettringite crystals were formed on the surface of the clinker grains, the distribution is inhomogeneous.

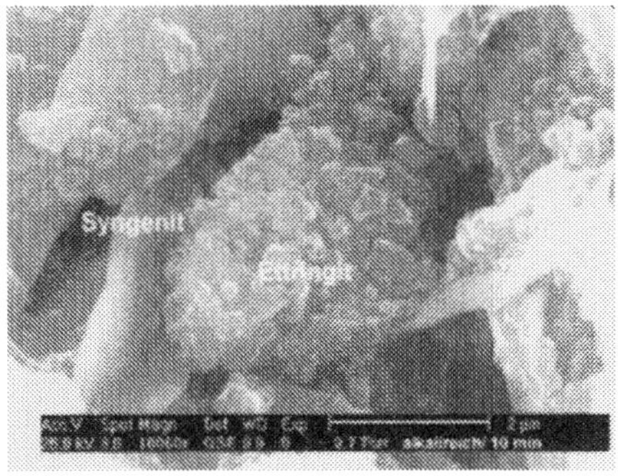

Figure 9b: Microstructure after 10 min hydration time [high potassium content]: Ettringite with the same morphology as cement with low arcanite content. Formation of syngenite (lath-shaped crystals).

The microstructure of cements with high potassium content (2 h) show ettringite and syngenite with the same morphology as after ten minutes (compare Figs. 9b and Fig. 10b).

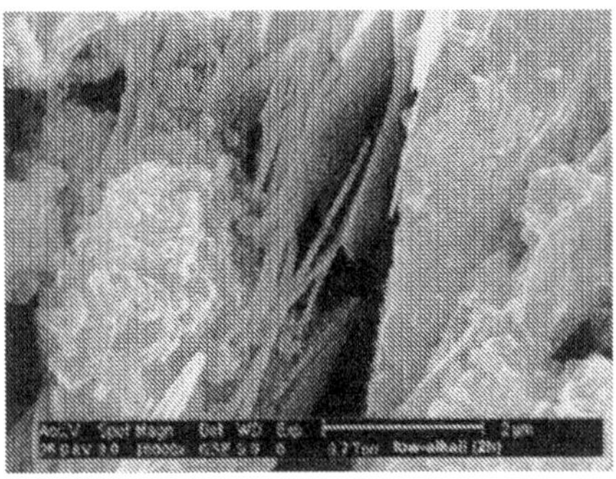

Figure 10a: Microstructure after 2 h hydration time [low potassium content]: Ettringite with the same morphology as after 10 min, increased syngenite formation.

Figure 10b: Microstructure after 2 h hydration time [high potassium content]: Ettringite and syngenite with the same morphology as after 10 minutes.

Acceleration period (high chemical activity): At the beginning of the acceleration period the C-S-H phases grow topochemical on the C_3S clinker surfaces. In the case of cements with low potassium content, small (< 200 nm long), isolated and highly inhomogeneously distributed C-S-H clusters can be observed after 3 h hydration time. Ettringite has nearly the same morphology as after ten minutes. After 5 h, the microstructure shows formations of small C-S-H phases with a length of up to 300 nm arranged in clusters (Fig. 11a). Individual clusters of fibers are formed. As the hydration continues further, the tipped needle like C-S-H phases grow longitudinally (one-dimensionally – up to 600 nm long in 10 h). Due to this one-dimensional growth, tight interlocking of the nanoscale C-S-H crystals occurs. The C-S-H phases formed in the hydration process of ordinary Portland cement during the first 24 hours exhibit a significantly faster growth than during the hydration of the pure C_3S clinker mineral. Up to 10 hours, the morphology of primary ettringite remains virtually unchanged. After a hydration period of 10 hours syngenite is no longer visible in the microstructure, but partial secondary gypsum is (Fig. 12a).

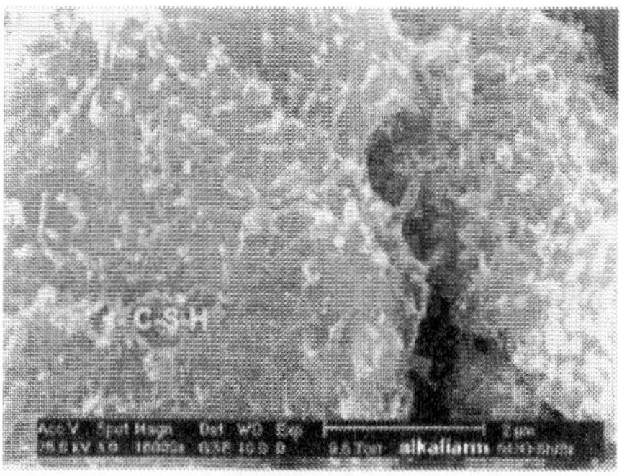

Figure 11a: Microstructure after 5 h hydration time [low potassium content]: Formation of small C-S-H phases (length up to 300 nm, arranged in clusters).

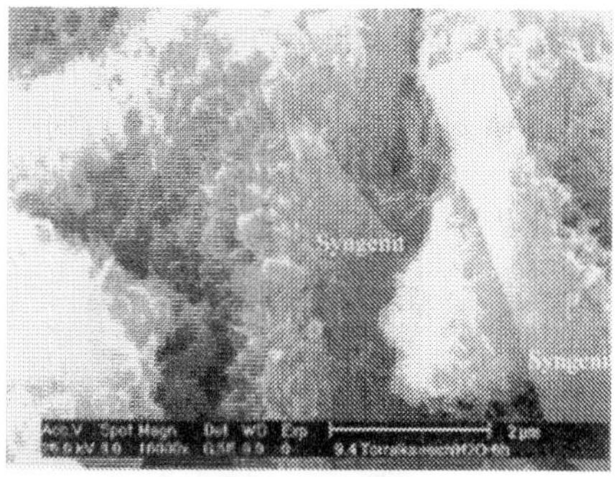

Figure 11b: Microstructure after 6 h hydration time [high potassium content]: syngenite is starting to dissolve with formation of secondary gypsum (left).

In cement with high potassium content, after 3 hours we observe many straight syngenite crystals in the form of stripes and plates and small but numerous C-S-H clusters. Ettringite has nearly the same morphology as after 10 minutes. Furthermore, the first portlandite crystals are visible.

After a hydration period of 5 hrs the lath shaped syngenite crystals start to dissolve and formation of secondary gypsum, potassium hydroxide and potassium sulfate (Fig. 11b) is observed. After a hydration period of 10 hours only a few syngenite crystals are visible in a matrix of C-S-H and ettringite (Fig. 12 b).

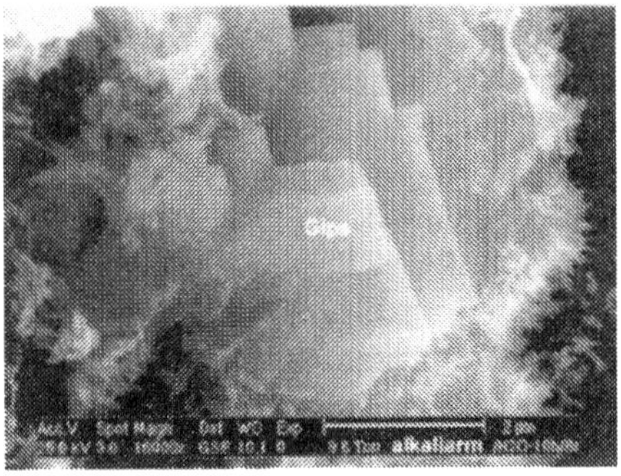

Figure 12a Microstructure after 10 h hydration time [low potassium content]: partial secondary gypsum, C-S-H (length up to 500 nm) and ettringite (length up to 600 nm).

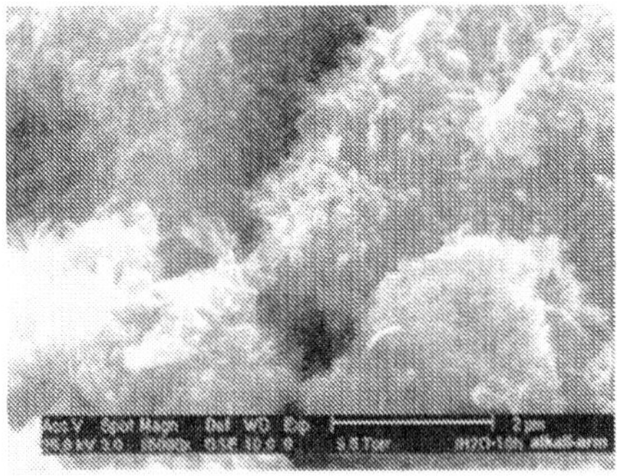

Figure 12b Microstructure after 10 h hydration time [high potassium content]: sporadic syngenite, C-S-H (length up to 500 nm) and ettringite (length up to 600 nm).

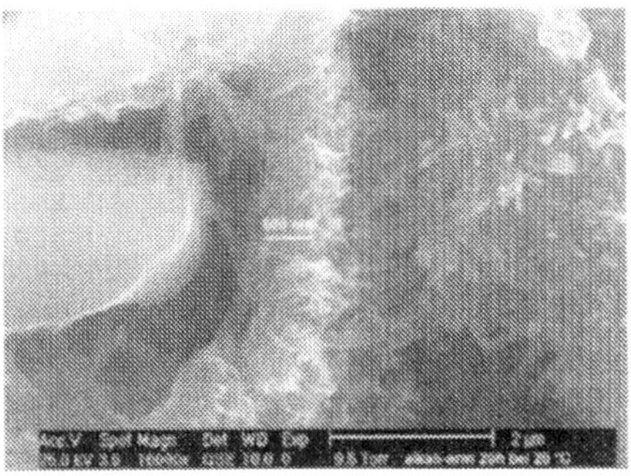

Figure 13a: Microstructure after 20 h hydration time [low potassium content]: ettringite crystal penetrate the space between the clinker grain and new formations of C-S-H.

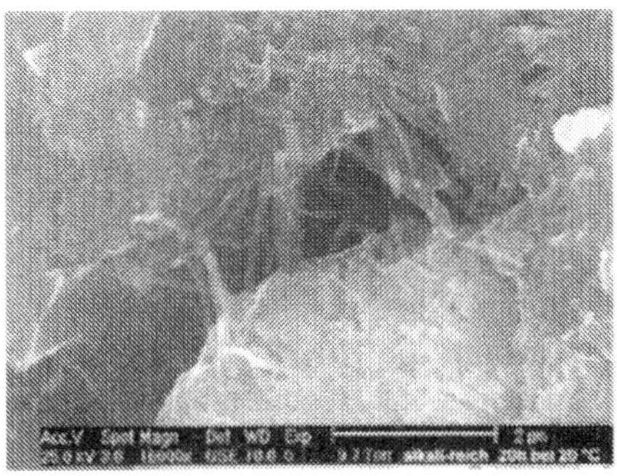

Figure 13b: Microstructure after 20 h hydration time [high potassium content]: C-S-H phases on the surface of an incompletely hydrated clinker grain (gap).

Deceleration period: Cement with low potassium content: The gap between the C-S-H reaction rim (about 600 nm thickness) and the hydrated clinker grain becomes even bigger (distance between 0.6 and 1 µm – Fig 13a). As the reaction process continues, the space is filled initially by ettringite and later also with C-S-H phases from adjacent clinker grains (Fig. 14a). The main growth phase of ettringite is between 12 and 24 hours. The length of the crystals is then up to 2.5 µm and they maintain their thickness all the time.

Cement with high potassium content: We also observe gaps between the hydration rim and the hydrated clinker grain. Ettringite crystals growing along their c-axis penetrate the gaps between the clinker grains and the hydration products (Fig. 13b). The longitudinal growth of ettringite can possibly be attributed to the newly released sulfate (dissolution of syngenite and formation of secondary gypsum). Further reason for the growth of ettringite is the release of C_3A through the progressive C_3S hydration. Calculations as well as measurements have shown that in the first hours only 10 % of the C_3A content can react. Sporadic thin hexagonal-plate calcium-monosulfoaluminate clusters are visible. Calcium-monosulfoaluminate is a solid solution of monosulfate and calcium-aluminate-hydrate. The voids and gaps were filled with C-S-H nanocrystals and ettringite (Fig. 14b).

Figure 14a: Microstructure after 24 h hydration time [low potassium content]: long prismatic ettringite (up to 2.5 µm), C-S-H (length approx. 600 nm).

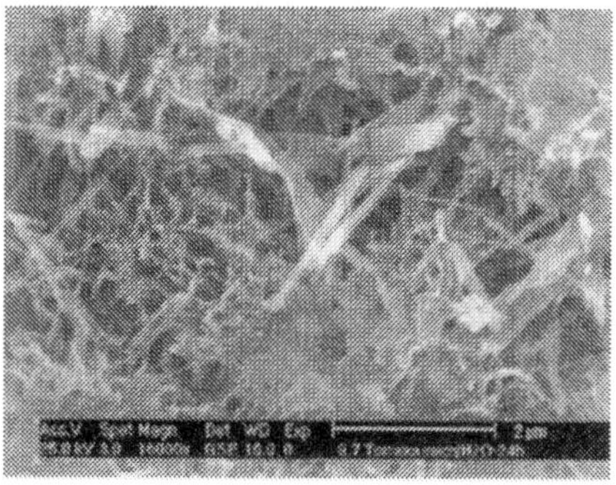

Figure 14b: Microstructure after 24 h hydration time [high potassium content]: voids in the microstructure are filled by C-S-H (length up to 600 nm) and ettringite.

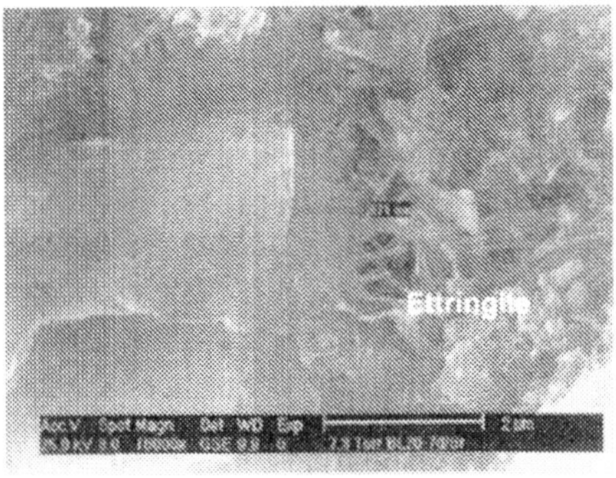

Figure 15a: Microstructure after 7 days hydration time [low potassium content]: long prismatic ettringite (up to 2.5 μm), and C-S-H (length approx. 900 nm).

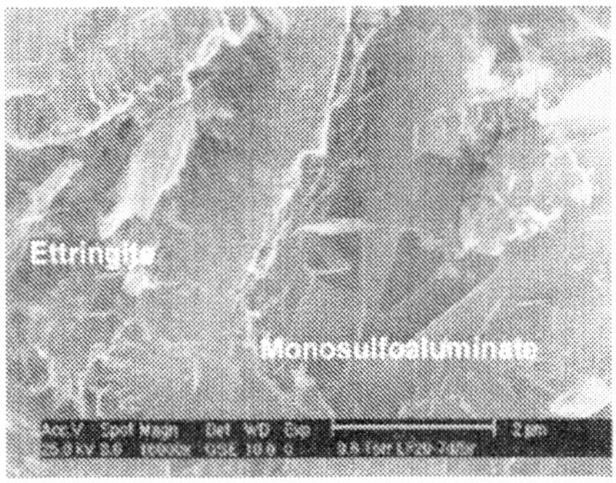

Figure 15b: Microstructure after 10 min hydration time [high potassium content]: voids of completely hydrated grains are filled, mainly with calcium-monosulfoaluminate.

Final period: During the final period there is further slow growth of the C-S-H phases. Additional basic consolidation of the microstructure occurs during this process, to which the slow reaction of the ß-C_2S also contributes.

After a hydration time of 7 days C-S-H reaction layers of about 900 nm thickness are visible. Figures 15a and b show typical occurrence of C-S-H reaction rims: 15a direct growth on the clinker grain, 15b C-S-H reaction rim around a completely dissolved clinker grain. Long prismatic ettringite (up to 2.5 μm) and thin hexagonal-plate calcium-monosulfoaluminate crystals are observable within the void of the clinker mineral (Fig. 15b). These completely hydrated clinker grains exist mainly in cements with a high soluble potassium sulfate content.

SCHEME FOR ORDINARY PORTLAND CEMENT HYDRATION

A scheme for hydration of OPC can be concluded from the above described observations. DCA and XRD measurements confirm this scheme. Figure 16 was derived to give a time resolved summary on the occurrence of hydration products.

In the first few hours, temporary syngenite formation is observed. The time of formation and the amount of the syngenite growth depend on the contents of soluble potassium sulfate (mainly arcanite). After a hydration period of about 5 hours the lath-shaped syngenite crystals dissolve and are replaced by secondary gypsum and potassium sulfate.

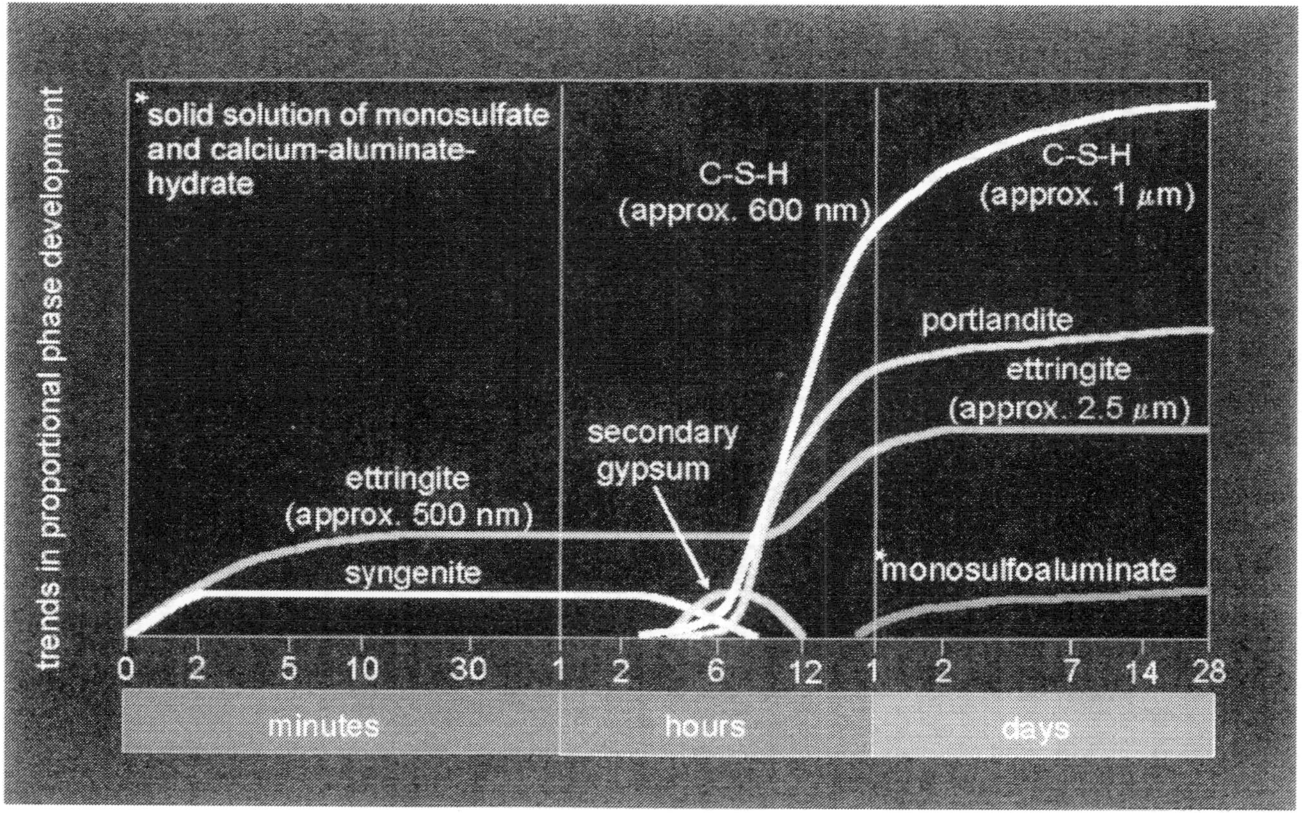

Figure: 16: Scheme for ordinary Portland cement hydration – derived from investigations by means of ESEM-FEG

Ettringite is the first hydration product formed shortly after mixing . The small prismatic ettringite crystals have a length of approximately 500 nm and a thickness of up to 250 nm. During the first hours the morphology of primary ettringite remains virtually unchanged. Minimal crystal growth is observed. The main longitudinal growth of these phases occurs between 12 and 24 h. The length of the crystals is then up to 2.5 μm and they keep their thickness all the time. Contrary to previous hydration models, ettringite does not become unstable and does not gradually convert into monosulfate or calcium-monosulfoaluminate. Both ettringite and calcium-monosulfoaluminate are the final products of hydration of OPC.

Calcium-monosulfoaluminate is formed, dependent on the content of water soluble potassium sulfate, after approximately 24 hours. Thin hexagonal-plate crystals are formed. At room temperature these crystals are solid solutions of monosulfate and calcium-aluminate-hydrate.

C-S-H phases form on the C_3S clinker phase at the beginning of the acceleration period. Small (about 200 nm long), isolated and strongly inhomogeneously distributed clusters of C-S-H phases are observed. Within 24 hours the C-S-H phases grow continuously to a length of 600 nm and the needle tips have a diameter of only approximately 5 nm. The longitudinal growth takes place from the surface of the clinker grain outwards. The result is a strong interlocking of the microstructure due to the extremely needle-like shapes.

The structure of μm-scaled portlandite crystal aggregates is determined by stacks of crystals with a thickness between 20 and 200 nm. This leads to large $Ca(OH)_2$ aggregates with a strongly inhomogeneous distribution.

HYDRATION OF OPC AT OTHER TEMPERATURES

The variation of OPC hydration with temperature is often discussed. Here we would like to give a short overview on our first insights of hydration of OPC at 5 °C and 62 °C.

OPC-Hydration at 5°C

At this temperature the thermodynamic stability (free reaction enthalpy) of ettringite and monosulfate is shifted to the stability of ettringite. After 21d hydration time ettringite crystals have a length of up to 4 μm (Fig.17a and b). Monosulfate or calcium-monosulfoaluminates are absent. Only pure calcium aluminate hydrates (C_4AH_{19}) are formed.

With increasing temperature the formation of ettringite decreases. At about 85 °C ettringite is completely unstable, i.e. it is not observable in the microstructure. In contrast, monosulfate is more stable than ettringite at higher temperatures. For example at a temperature of 62 °C, in addition to ettringite, monosulfate, hydrogarnet and hydrotalcite can be found (Fig.18a and b).

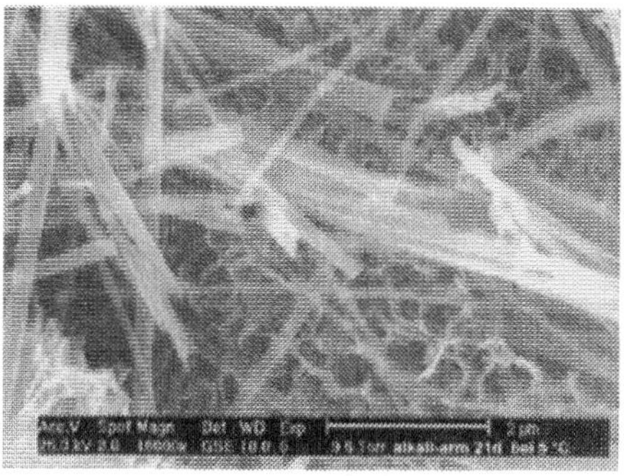

Figure 17a: Microstructure after 21 days hydration time at 5 °C: very long prismatic ettringite (up to 4 μm).

Figure 17b: Microstructure after 21 days hydration time at 5 °C: hexagonal calcium-aluminate-hydrate.

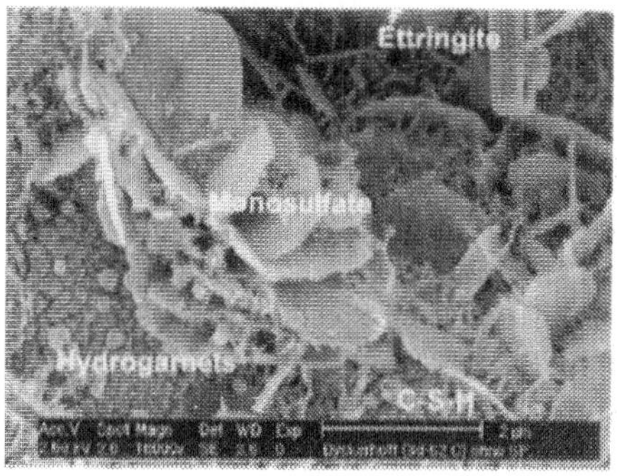

Figure 18a: OPC after 3 days at 62 °C: area of a completely hydrated clinker grain: hydrogarnet: solid solution of C_3AH, C_3AS_3, C_3FH_6 and C_3FS_3: cubic.

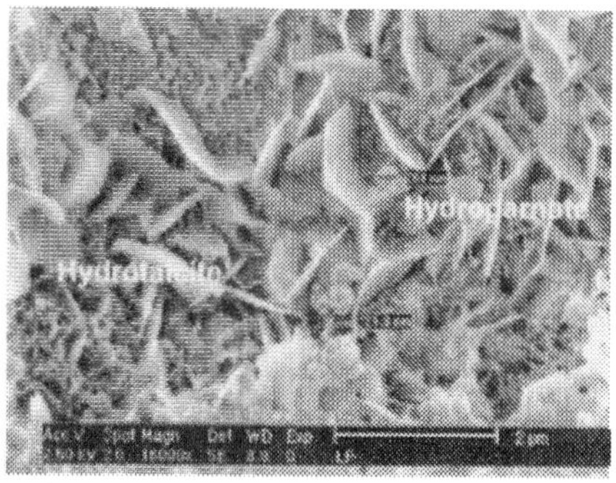

Figure 18b: OPC after 3 days at 62 °C: area of a completely hydrated clinker grain: hydrogarnet and hydrotalcite (thin plates) $Mg_6Al[CO_3(OH)_{16}]4H_2O$.

All of these phases are mainly found in areas of completely dissolved clinker grains. Hydrogarnets - a solid solution of C_3AH, C_3AS_3, C_3FH_6 and C_3FS_3 – are cubic crystals (approximately 250 nm) whereas hydrotalcite - $Mg_6Al[CO_3(OH)_{16}]4H_2O$ – are crystallizes in the form of thin (approximately 25 nm thick) plates. At room temperature monosulfate is less stable than ettringite.

This can be illustrated by the following experiment: monosulfate and water were stored at 22 °C in a glove box. After a relatively short period of time, monosulfate dissolves and the formation of hexagonal-plate calcium-aluminate-hydrates as well as ettringite is visible (Fig. 19a and 19b).

Figure 19a: Storage of monosulfate in water at 22 °C: dissolution of monosulfate and formation of ettringite and calcium-aluminate-hydrate.

Figure 19b: Storage of monosulfate in water at 22 °C: formation of aggregates of parallel grown ettringite crystals.

CONCLUSIONS

It has been shown that the Environmental Scanning Electron Microscope offers distinct benefits over a conventional SEM for the observation of hydrate phases. The hydration process can be followed realistically and the structures can be reproducibly imaged. It is possible to observe images of the temperature, and vacuum, sensitive and only several nanometer-thick hydration products of OPC. New insights into the early hydration state can especially be obtained. Sample preparation artifacts are very limited.

Comparative studies using the ESEM-FEG in different operating modes, for example, wet- (ESEM mode) and high vacuum- (conventional SEM) mode have been carried out. It can be shown that with the traditional methods (drying, coating, beam damage) changes in the morphology of the C-S-H phases are dramatic.

Furthermore, the article describes investigations of phase development during the hydration of OPC from the induction to the final period. The results are summarised in a scheme for the normal hardening process.

The comparison of the ESEM investigations with DC analyses reveals that the typical two maxima in these DCA-curves give a good correlation with microstructure development. The second maximum of the DCA-curves (acceleration period) corresponds with the first appearance of C-S-H phases.

REFERENCES

1. Powers, T.C.,Helmuth, R.A. (1953) Theory of Volume Changes in Hardened Portland Cement Paste during Freezing , Proc. Highway Research Board 32, 285
2. Powers, T.C. (1960) Pysical Properties of Cement Paste, Proc. 4. Int. Symposium on the Chemistries of Cement, Washington, Vol. II, S.577
3. Feldmann, R.F., Sereda, P.J. A Model for Hydrated Portland Cement Paste as deduced from Sorption length Change and mechanical Properties (1968), Materials and Structures 1, 509
4. Kondo, R., Daimon, M. Phase Composition of hardenet Cement Paste (1974), auf dem VI Int. Congress on the Chemistry of Cement in Moskau
5. Richartz, W., Locher, F.W. Ein Beitrag zur Morphologie und Wasserbindung von Calciumsilicathydraten und zum Gefüge des Zementsteins (1965), Zement Kalk Gips 18, S. 449 – 459
6. Locher, F.W.; Richartz, W.; Sprung, S. Erstarren von Zement Teil 1: Reaktion und Gefügeentwicklung (1976), Zement Kalk Gips 29, S. 435 - 442
7. Locher, F.W.; Richartz, W.; Sprung, S. Erstarren von Zement Teil 2: Einfluß des Calciumsulfatzusatzes (1980), Zement Kalk Gips 33, S. 271 - 277
8. Locher, F.W.; Richartz, W.; Sprung, S.; Sylla, H.-M. Erstarren von Zement Teil 3: Einfluß der Klinkerherstellung (1982), Zement Kalk Gips 35, S. 669 - 676
9. Locher, F.W.; Richartz, W.; Sprung, S.; Rechenberg, W. Erstarren von Zement Teil 4: Einfluß der Lösungszusammensetzung (1983), Zement Kalk Gips 36, S. 224 - 231
10. Setzer, M.J. Einfluß des Wassergehalts auf die Eigenschaften des erhärteten Betons (1977), Dt. Ausschuß für Stahlbeton Heft 280
11. Wittmann, F.H. Grundlagen eines Modells zur Beschreibung charakteristischer Eigenschaften von Beton (1977), Dt. Ausschuß für Stahlbeton Heft 290
12. Taylor, H.F.W. Cement Chemistry (1992) Academic Press Limited San Diego, Second Printing
13. Massazza, F., Daimon, M. Chemistry of Hydration of Cements and Cementitous Systems (1992) IX Int. Congress on the Chemistry of Cement in New Dehli, Congress Reports-Vol.1 S. 383 – 448
14. Meredith, P., Donald, A., Luke, K. Pre-induction and Induction Hydration of Tricalcium Silicate: An Enviromental Scanning Electron Microscopy Study (1995) Journal of Matreial Science 30, Cambrige , S.1921 - 1930
15. Jennings, H.M. A Model for the mocrostructure of calcium silicate hydrate in cement paste (2000) Cement and Concrete Research 30, S. 101 - 116
16. Mindess, S., Young, J. F. Concrete (1981) Prentice-Hall, New Jersey
17. Stark, J., Möser, B., Eckart, A. New approaches to cement hydration (2001) ZKG International 54, S. 52 – 60 and 114 - 119
18. Hader, K. Untersuchung der Zementhydratation im frühen Erhärtungsstadium mittels ESEM-FEG, Diplomarbeit 2000 an der Bauhaus-Universität Weimar

Photo 27: B. Möser

Photo 28: A. Brandt and S. Shah

NANOTECHNOLOGY AND CONCRETE: WHAT CAN WE UTILISE FROM THE UPCOMING TECHNOLOGIES?

Pavel Trtik and Peter Bartos
Scottish Centre for Nanotechnology in Construction Materials
ACM Centre – Nanocom, University of Paisley
High St, Paisley, PA1 2BE
Scotland, UK

ABSTRACT

Recent progress in research and development in the field of assessment of nano/micromechanical properties of materials has demonstrated very high potential for substantial improvements in their design and composition. The paper will focus on the use of (a) nanotechnology-based testing techniques, and (b) nanotechnology-based tools for microfabrication in research and development of cementitious materials.

Depth sensing indentation (DSI) test methods (such as micro/nanoindentation) are becoming increasingly important tools for the characterization of the mechanical properties of various materials. During the DSI test procedure a probe/indenter is impressed into the material tested while both the load imposed on the testing site and the displacement are continuously monitored. As a result, an impression - micro/nano indent - is left on the surface of the specimen. The common properties, such as hardness, elastic modulus and/or fracture toughness of bulk materials can consequently be derived from the displacement/load diagram. Also, DSI test methods can be utilized for assessment of interfacial properties of fiber reinforcement.

Concerning nanoscale-precision microfabrication, the paper will pay particular attention to focused ion beam (FIB) techniques and their use for cement-based materials. FIB techniques include nanoscale-precision material milling and material deposition. Further potential applications of nanotechnology-based processes in construction materials are discussed.

INTRODUCTION

Nanotechnology is a very fast developing area of research. In its broadest definition, nanotechnology is viewed as 'the technology where dimensions or tolerances in the range of 0.1 to 100nm play a critical role'. It is then not surprising that the term nanotechnology means different things to different people. Nanotechnology is not a homogeneous field; it subsumes a wide range of technological and scientific developments on nanoscale.

There are two distinct types of nanotechnology. 'Top-down' nanotechnology refers to the fabrication of nanoscale structures by machining and etching techniques, whereas 'bottom-up', often referred to as molecular nanotechnology, applies to the creation of organic and inorganic structures, atom-by-atom, or molecule-by-molecule.

Nanotechnology has already had a significant impact on many aspects of life and the construction industry, in spite of its conservative nature, will not be omitted from the 21st century industrial revolution. Despite all recent developments, an era in which structural elements of high-rise buildings, bridges, etc. will be constructed from 'close-to-ideal' materials produced in molecular assemblers - by the 'bottom-up' approach i.e. atom by atom - is still, unfortunately, very distant from now. Before the advent of such times, it is necessary to utilize advances in the 'top-down' approach of nanotechnology and in nanotechnology-based testing techniques in order to acquire a much better understanding of the behavior of "conventional" construction materials, such as concrete. Such enhanced understanding of processes in complex construction materials, translated into much more economical use of materials and possibly also leading to construction materials of higher durability (thus, to reduced costs in the repairs/maintenance of structures), is the ultimate goal of the current research.

Two specific domains of nanotechnology applications for research and development of concrete and other cementitious materials are identified: (i) nanotechnology-based testing techniques for assessment of mechanical properties in nanometre regime and (ii) nanoscale-precision microfabrication techniques. The significance of the extension of nanotechnology-based testing techniques and nanoscale fabrication processes from the materials science domain to the area of 'real' construction materials cannot be overstated as such materials are ubiquitous and used in vast quantities. Concrete is the second most widely used commodity in the world after only water and approximately 1 cubic metre (2.5 tonnes) of concrete per capita is cast in Europe every year. Therefore, envisaged advances in the performance of cementitious materials (even if only incremental), stemming from a better understanding and control over their properties assessed in the micro/nanoscale, may have major economic and environmental impacts.

TESTING TECHNIQUES FOR ASSESSMENT OF MECHANICAL PROPERTIES OF CONCRETE IN THE NANO/MICROSCALE

Extensive research and development in the field of testing of nano/micromechanical properties of materials have demonstrated very high potential for substantial improvements in design and performance of many materials. The progress in testing of mechanical properties in nanoscale has been, in particular, stimulated by the demand for assessment of the properties of thin coatings. Taking into account the often submicron thickness of such films, the use of nanotechnology-based testing techniques became apparent as the only possible method for quantitative testing of their mechanical properties.

However, the use of nanotechnology-based testing methods has not remained confined within the area of thin coatings. Presently, useful applications of these testing techniques can be found in other areas and is leading to the development of better new generation materials.

The nanotechnology-based testing apparatuses for assessment of mechanical properties - mechanical property nanoprobes (MPN) - can be defined as tools capable of a continuous monitoring of both load and displacement with respective sub-nanometre and nanoNewton resolutions. These tools include surface force apparatuses (SFA), atomic force microscopes (AFM) and nanoindenters. The following parts of this chapter will focus on the utilization of nanoindentation – often called depth-sensing indentation (DSI) or instrumented indentation - for assessment of the mechanical properties of concrete and other cementitious materials.

Indentation Testing Techniques

Microhardness Tests: Microhardness test methods are the direct predecessors of nanotechnology-based indentation techniques. There are several types of hardness test, such as Vickers, Brinell and Rockwell hardness tests. In these - conventional - hardness tests, the size of the indent is measured after the load is removed from the test site and therefore the 'residual' hardness is derived. The hardness value is the only output of such tests and the bottom limit of the indentation size is given by the resolution of imaging.

The utilization of hardness tests for assessment of cementitious materials has mainly been focused on characterization of the interfacial transition zone (ITZ) around various inclusions (aggregate, fibers, etc.) by the Vickers or Knoop microhardness test. The first reported use of hardness testing of the interfacial zone[1] goes back to 1962 and since then an extensive number of publications can be found in the literature. Apart from the hardness testing of ITZ, several investigations[2] have been reported so far into the use of microhardness testing for 'non-destructive' assessment of the properties of bulk plain cement pastes. This research provided correlation of the microhardness values with macromechanical properties, such as compressive strength, Young's modulus, etc.

However, one of the surprising findings of a literature review[3] has been the apparent lack of detailed images of real indents (see Figure 1) in the publications reviewed. Also, hardly any of the reviewed publications reported information on the loading rates and the dwell times employed during the testing procedure.

Figure 1- SEM image of an indent in cement paste.
The scale bar corresponds to 2 μm.

Depth Sensing Indentation: In 1983, Pethica[4] reported the development of a depth sensing indentation facility with the respective micro-Newton and nanometre range. Figure 2 shows a load vs. displacement diagram – a typical outcome of such a DSI test. Continuous recording of load and depth during the test now allows determination of not only hardness but also many other key material parameters such as indentation elastic modulus, indentation creep, etc. Also, elastic and plastic work can be readily derived from the load vs. displacement diagram. As a direct consequence, DSI testing allows assessment of the elastic modulus of the ITZ and its comparison with the elastic modulus of "bulk matrix". The authors of this paper carried out such investigations and the outcome of the testing program is reported elsewhere.[5] The depth sensing indentation has also recently been used by Velez et al.[6] for assessment of pure constituents of Portland cement clinker.

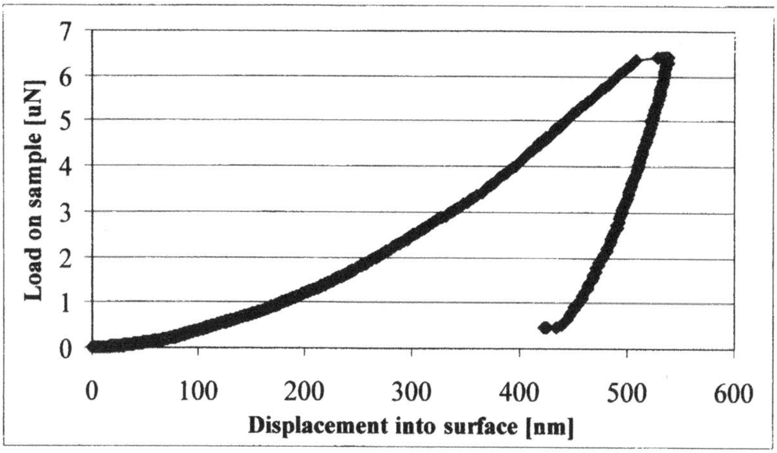

Figure 2 - Load vs. displacement diagram of depth-sensing indentation test performed on sample of cement paste. The displacement at the end of the loading part corresponds to 500 nm.

Continuous Stiffness Method (CSM-DSI): The technique introduced by Pethica and Oliver[7] in 1987 improves the quasi-static DSI testing instrument to a dynamic measurement instrument. The CSM technique imposes an oscillating force (with a frequency of usually about 50 Hz) of very low amplitude (as low as a tenth of μN) on the indenter during the loading part of the DSI test procedure. CSM systems are able to record displacement amplitudes - induced by excitation force oscillation - as small as 1 Å. Consequently, the phase difference between the oscillation of excitation force and displacement signals is evaluated. The testing instrument is modeled as a dynamic system. Therefore, each indentation instrument has a slightly different rheological model. [8] The contact stiffness and consequently elastic modulus and hardness can thus be determined continuously for each individual indentation as a function of displacement (see Figure 3).

Limitations of DSI Testing Methods

With the introduction of the CSM technique that allows the continuous determination of mechanical properties from the very start of the test (i.e. in very shallow depths of the indentation process) the surface roughness and, hence, sample preparation procedure must be seriously considered. The indentation depth must be large enough with respect to the surface roughness. The surface roughness can be readily investigated by atomic force microscopy. Figure 4 shows 3D AFM map of an indent.

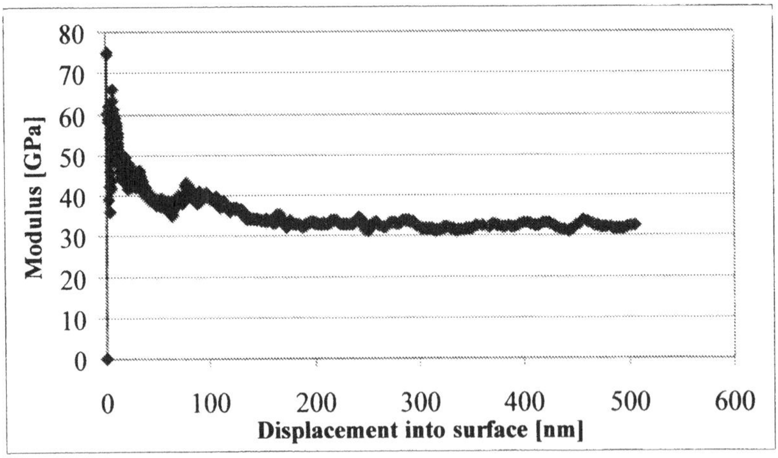

Figure 3 - Indentation elastic modulus monitored continuously as a function of displacement into surface during CSM-DSI test

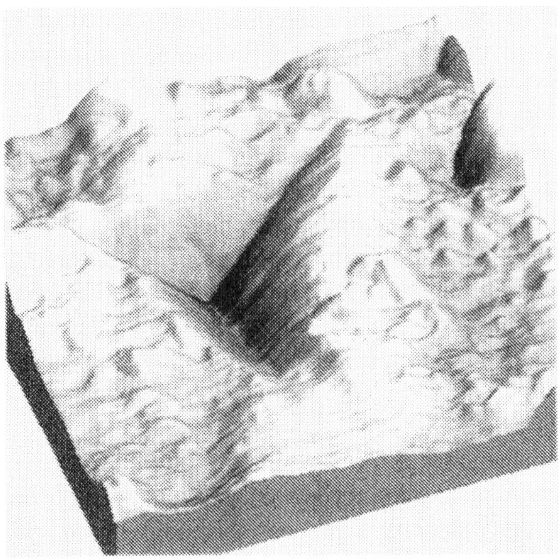

Figure 4 - 45°-tilted 3D AFM image of 500nm-deep indent in cement paste. Size of the scanned area is 8.3 by 8.3 μm.

The issue of surface preparation becomes most apparent for nanoindents located in very close proximity to the aggregate/cement paste interface. Due to the usually higher hardness the surface of the aggregate protrudes above the surface of the cement paste about a few microns. Figure 5highlights the necessity of post-test observation of the indentation site in order to confirm validity of the test.

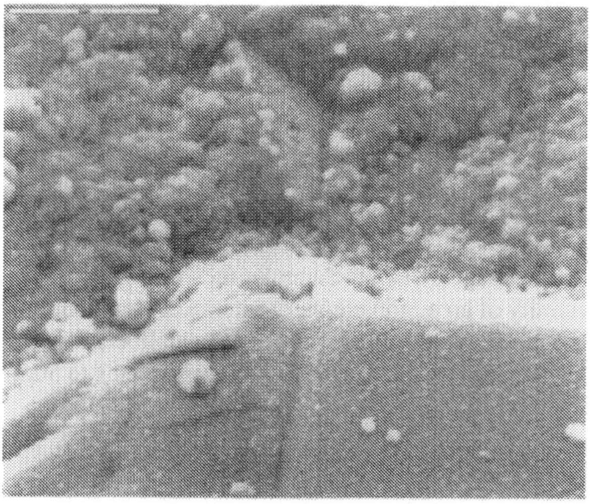

Figure 5 - SEM image of an indentation site in very close proximity to an aggregate/cement paste interface. Note the impression of the edge of the indenter in the aggregate.

Other issues that must be taken into account in DSI test methods include the thermal stability of the testing environment and the correct calibration of the tip shape function.

Future Prospects of DSI Testing Techniques in the Domain of Cementitious Materials

With the current sub-micron positioning accuracy of nanoindenters, site-specific nanoindentations can be easily carried out in a pre-programmed manner. The positioning capabilities of modern nanoindenters are demonstrated in Figure 6. The logical direction of future research is therefore aimed at the correlation of nanomechanical testing with assessment of chemical composition of the indentation test sites by back scattered scanning electron microscopy (see Figure 7) and/or quantitative chemical analysis.

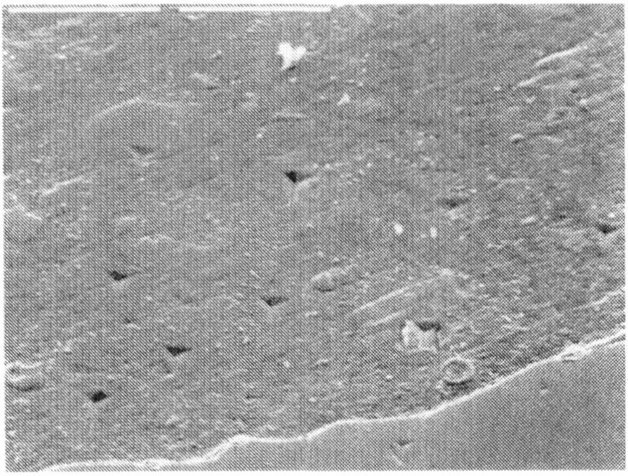

Figure 6 - SEM image of 4x3 rectangular mesh of 700nm-deep indents carried out in the vicinity of an aggregate/paste interface. Note that the bottom right indent in the mesh was pre-programmed to be deliberately carried out in a different position. The scale bar corresponds to 50μm.

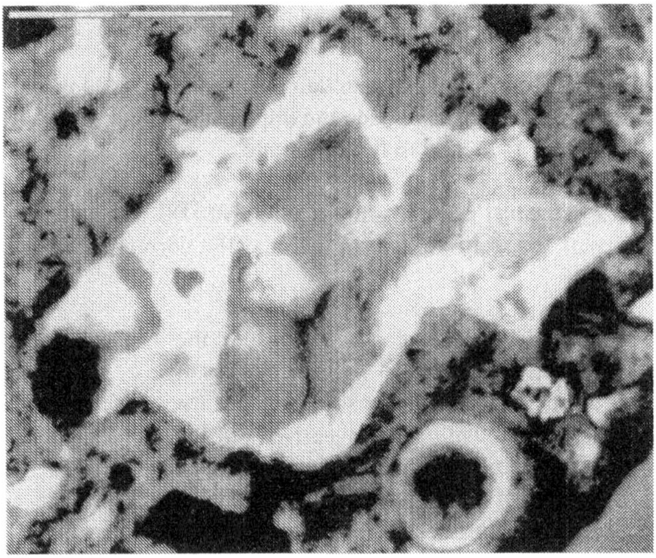

Figure 7 - Detailed BS-SEM image an indent (one of the indents shown in Figure 6) and its surrounding area.

NANOSCALE-PRECISION MICROFABRICATION TECHNIQUES

Focused ion beam milling/etching and focused ion beam deposition represent typical 'top-down' nanoscale fabrication techniques. The principal area of utilization of focused ion beams (FIB) is well established in microelectronics, namely in the semiconductor and integrated circuits industries, where ion beams are commonly used for the modification of prototype integrated circuits.

Focused ion beams (FIB) are generated by metal field emission ion sources. The liquid metal ion source (LMIS) providing heavy metal ion beams was developed in the mid-1970s, thus starting off a considerable growth in FIB applications. The majority of FIB systems employ gallium ions (Ga^{+}) accelerated to 20-30 keV, although ions can be produced with liquid-metal ion sources from other elements, such as Al, As, Au, B, Be, Bi, Cs, Cu, Ge, Fe, In, Li, P, Pb, Pd, Si, and Zn. Ion beams are focused, by electrostatic lenses, to dimensions as small as about 10 nm.

The use of FIBs for research in the domain of cementitious materials can be subdivided into two distinct areas:

Post-Indentation Observation of Test Sites

Focused ion beam etching/milling processes can also be used for advanced microstructural examination of nanoindents. Until now, a nanoindent could only have been observed from 'above', as an impression into the surface of the tested material. In its cross-sectioning mode, FIB milling enables the material to be etched out in such a manner that the previously unavailable but very important observation of microstructure of the material underneath the indent can be carried out.

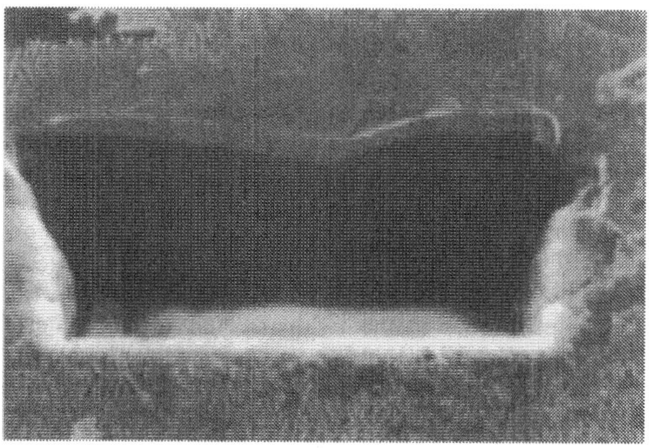

Figure 8 - FIB image of the cross-section of an indent in cement paste.

As the cross sectioning reveals entirely new information about the 'third dimension' of the microindents and about the microstructure of the material surrounding the indent, these novel observations can then be related to the displacement vs. load diagrams obtained during the indentation experiment and help in their interpretation. Important subsurface features such as nanoindentation-induced microcracks can be observed[9]. However, due to its insulating nature, concrete is not quite suitable for direct observations of such subsurface features in FIB. For the best image to be taken, the etched surfaces of the sample should be recoated and observed in field emission SEM (see and compare Figures 8 and 9).

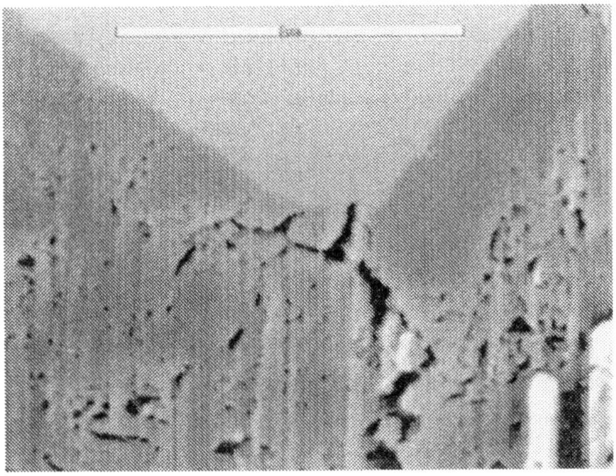

Figure 9 - Detailed field emission SEM image of the cross-section of an indent in cement paste. Scale bar corresponds to 5μm.

Also, the time-consuming multiple 'slicing' of the indentation site by FIB etching in order to provide 3D microstructure mapping of indentation sites (suggested as a possible experimental technique by the authors[3]) was recently reported by Inkson[10] in a study of subsurface deformation of CU-Al multilayers.

Pre-Indentation Sample Preparation

Hitherto, the authors are not aware of any application of FIBs for sample preparation of DSI techniques in the field of cementitious materials. However, Wissing et al[11] utilized glancing incidence ion beams in the preparation of nearly atomically flat surfaces. FIB etching techniques can be used for tailored preparation of specimens to be assessed by DSI test methods. Various three-dimensional objects, such as beams, cantilevers etc., can be micromachined in the structure of the specimen to be indented. Such a microbeam or microcantilever is then loaded in the indentation apparatus.

Elastic and plastic properties of materials can thus be investigated from the deflection of such microbeam/microcantilever[12]. In this particular example of the DSI technique, a spherical, conical or flat-ended indenter should be used in order to minimize the size of the indent on the top surface of the microbeam or to prevent local fracture of the loaded place of the microbeam. The mechanical properties can be derived from the displacement (deflection of the cantilever)/load diagrams in the same manner as if they were determined in macroscale. Since the dimensions of the micromachined object can be measured with a high precision and the deflection vs. load diagram is monitored during the 'indentation' experiment, the elastic modulus of the material of the microbeam/microcantilever can be derived from the classical bending theory of a beam. However, there are hardly any reports available on the influence of ion beam treatment on the near-surface mechanical properties of the material tested.

CONCLUSIONS

The paper summarizes the advances in testing techniques for the assessment of mechanical properties of cementitious composites in the nano/microscale and in the utilization of nanoscale-precision microfabrication techniques.

Depth-sensing indentation test methods provide a powerful tool for investigation into the mechanical properties of cementitious materials in nanoscale. Apart from hardness, which can also be assessed by the Vickers microhardness test, indentation elastic modulus and indentation creep can also be derived from the nano/microindentation test. The correlation of the nanoindentation test results with quantitative chemical analysis may provide very useful information about the mechanical properties of basic constituents of cementitious materials.

The nanoscale precision fabrication methods, represented by focused ion beam milling/etching and focused ion beam deposition techniques show large potential for a better interpretation of the nanoindentation test and an improved understanding of the microfracture processes in cement-based materials.

REFERENCES

[1] T. Lyubimova and E.R. Pinus, "Crystallisation structure in the contact zone between aggregate and cement in concrete", *Kolloidnyi Zhurnal (USSR),* **24** 578-87 (1962)

[2] F.R. Feldman and H. Cheng-yi, "Properties of Portland cement-silica fume pastes, II. Mechanical properties", *Cement & Concrete Research,* **15** 943-952 (1985)

[3] P. Trtik, "Aspects of micromechanical properties of cement-based materials", Ph.D thesis, University of Paisley, Scotland, UK, December 2000, pp. 204

[4] J.B. Pethica et al., "Hardness measurements at penetration depths as small as 20 nm", *Philosophical Magazine A,* **48** 593-606 (1983)

[5] P. Trtik and P.J.M. Bartos, "Micromechanical properties of cementitious composites", *Materials & Structures*, **32** 388-393 (1999)

[6] K. Velez et al., "Determination by nanoindentation of elastic moduli and hardness of pure constituents of Portland cement clinker" *Cement & Concrete Research,* **31** 555-561 (2001)

[7] J.B. Pethica and W.C. Oliver, "Tip surface interactions in STM and AFM", *Physica Scripta*, **T19** 61-67 (1987)

[8] N. Burnham, S.P. Baker and H.M. Pollock, "Model for mechanical properties nanoprobes", *Journal of Materials Research*, **15** 2006-2014 (2000)

[9] P. Trtik, C.M. Reeves and P.J.M. Bartos, "Use of focused ion beam (FIB) for advanced interpretation of microindentation test results applied to cementitious composites" *Materials & Structures*, **33** 189-193 (2000)

[10] B.J. Inkson et al. "Subsurface nanoindentation deformation of CU-AL multilayers mapped in 3D by focuseed ion beam microscopy", *Journal of Microscopy*, **201** [2] 256-259 (2001)

[11] M. Wissing, M. Batzill, and K.J. Snowdon, "Preparation by glancing incidence ion irradiation of CaF2 surfaces with angstrom-scale RMS roughness", *Nanotechnology*, **8** [1] 40-45 (1997)

[12] O. Kraft, R. Schwaiger, W.D., 'Measurement of mechanical properties in small dimensions by microbeam deflection', *Materials Research Society Symposium Proceedings*, **518** (1998) 39-44

Photo 29: P. Trtik

Photo 30: B. Mitchell, J. Skalny, D. Mitchell

Photo 31: A. Bajza, E. Bajza and D. Turk

Photo 32: J. Lukasik, A. Olek, A. Brandt and J. Olek

FUTURE GOALS OF CEMENT AND CEMENTITIOUS MATERIALS FOR HIGH PERFORMANCE CONCRETE

Michael Bergin, State Structural Materials Engineer
Ghulam Mujtaba, State Prestressed Concrete Materials Engineer
Charles Ishee, Structural Materials Engineer
Florida Department of Transportation
2006 Northeast Waldo Road
Gainesville, FL, USA, 32609

ABSTRACT

The Florida Department of Transportation (FDOT) has developed concrete and concrete materials related specifications and guidelines with the intention that they provide maximum durability and protection to the Department's bridge and roadway structures. These protective measures include the environmental classifications of roadway and bridge sites, use of concrete in accordance with its environmental conditions, minimum concrete cover, maximum allowable chloride content, use of pozzolanic materials, blast furnace slag, and corrosion inhibitors.

Special mixes are used and special measures are taken to provide the required plastic and mechanical properties especially in hot weather concrete, flowing concrete admixtures and self-leveling concrete. The uses of environmentally safe and locally available materials (such as petroleum coke fly ash, bark ash, Class F fly ash, silica fume, calcium nitrate, ground granulated blast furnace slag, and metakaolin) are allowed as an ingredient in the concrete when proven to provide the required durability and strength.

The future concrete related aim and mission of the FDOT are the optimization of concrete material ingredients with their required performance. The development of performance-based specifications and criteria are needed. A thorough study of the modulus of elasticity, creep, shrinkage, and other concrete properties for locally available materials should be studied. The use of non-destructive testing for field performance evaluation should be increased with the intent of utilizing this equipment for acceptance of the concrete. The precast/prestressed concrete, drainage products and ready-mixed concrete plants should meet the requirements of national standards as eligibility requirements for providing concrete or concrete materials to FDOT projects. The FDOT, suppliers, and manufacturers of concrete materials, as well as academic research centers, need to work together to accomplish the future concrete related aim and mission of the State of Florida.

INTRODUCTION

In order to project the trends and to elaborate on the challenges of concrete and cementitious materials, this paper summarizes a brief review of concrete related specifications. In addition, the report provides current research activities and proposed specification changes to enhance the durability of concrete bridges and roadway structures. The last portion of the report discusses the future research and materials testing needs for the future of high performance concrete.

Currently, Florida maintains more than 6000 bridges along the longest coastline in the union; this includes over 100 bascule bridges. One familiar bridge is the Sunshine Skyway Bridge across Tampa Bay. It is approximately 7 miles long and utilized 225,000 cubic yards, almost a million tons, of concrete. The aggressive nature of Florida's environments include the effects of intercoastal waterways and tidal effects of the Atlantic and Gulf waters. Structures constructed of concrete require special ingredients to withstand the destructive environmental forces. As a result, the Department is very particular in the selection of materials allowed for use in the construction of bridges and other structures. FDOT has been specifying the use of customized concrete mixes and curing practices based on the strength and environmental classifications of the projects. The bridge project sites are classified as slightly aggressive, moderately aggressive and extremely aggressive. The purpose of these different requirements is to increase durability and service life and decrease the future costs that come with rehabilitation of these structures. The cost of repairs on structures and construction are becoming an increasing burden on the taxpayer. In addition to the monetary requirements, the motoring public equates the process of construction to traveling delays and an increasing nuisance.

With this in mind, FDOT has developed concrete and concrete material specifications and guidelines with the intention that they provide maximum durability and protection to the Department's bridges and roadway systems. These protective measures include the environmental classification of roadway and bridge sites, use of concrete in accordance with its environmental conditions, minimum concrete cover, reduced allowable chloride content and the use of pozzolanic materials, blast furnace slag and corrosion inhibitors.

SPECIFICATIONS

Based on previous research, the Department has developed specifications to ensure protection of our structures. Specific concrete design mixes are required depending on the environmental conditions of the project site. The environmental conditions include; “slightly”, “moderately”, and “severely” aggressive, based on the pH, resistivity, sulfates, and chloride content of the soil or water.

For the purposes of this workshop, the focus will be on the issues concerning the requirements to ensure protection in aggressive environments. Concrete design mixes are developed to ensure protection of reinforcing steel in structures constructed in severely aggressive environments. Design mixes for this environmental classification typically include the following:

Cement is required to meet AASHTO M-85 Type II. This mandates low alkali cement with a C_3S limit of 58% and the C_3A restricted to 8% maximum. The total cement content is based on the class of concrete for the intended application.

Pozzolans are replaced on a pound per pound basis with cement and may include Class F fly ash or ground granulated blast furnace slag (GGBFS), silica fume, or metakaolin.

Dosages of theses Pozzolans or Slag include:

- Fly Ash in an 18 to 50% replacement in mass concrete, 33 to 37% replacement in drilled shafts. All other concretes require an 18 to 22 % replacement.
- Silica Fume is replaced at 7 to 9% by weight of the cement and is required when the environmental classification is extremely aggressive due to the presence of chlorides in water. Silica fume is specified in the splash zone of structures. The splash zone is considered to be 12 feet in the vertical direction, measured as the distance between the lowest superstructure elevation and the water surface at Mean High Water (MHW).
- Metakaolin is required in the same locations as silica fume and is replaced at 8 to12% of the cement mass.
- GGBFS can be substituted for fly ash at the following replacement rates, 58 to 62% in drilled shafts, 50 to 55% when used with silica fume or metakaolin, and 50 to 70% in Mass Concrete and all other applications.

Water to Cementitious Materials Ratio for substructure (typically piles) utilizing silica fume is 0.35. These low water binder ratios require a high range water reducing admixture. In addition, the chloride permeability of the concrete is required to be less that 1000 coulombs at 28 days, when tested in accordance with AASHTO T-277.

Corrosion Inhibitor is allowed and must be used with Type II cement and fly ash or slag. In addition, a water reducing and retarding admixture is required.

Reinforcing Steel will have three inches of clear cover for adequate protection in a severely aggressive environment. Epoxy coated rebar is not allowed for use in any FDOT construction project and the use of stainless steel clad rebar is currently under review.

Chloride Content - the maximum allowable chloride content of prestressed concrete during production is required to be less than 0.40 lbs/yd^3 and less than 0.34 lbs/yd^3 during laboratory trial batching.

SPECIAL MEASURES

Florida experiences temperature variations that require additional procedures for both hot and cold weather. Due to the close proximity of construction projects to Gulf of Mexico or Atlantic sea waters, the humidity is high most of the time. This high humidity has assisted the concrete industry to furnish and place concrete during hot weather. Special mix designs in Florida may be approved for use in hot weather conditions. This means that concrete can be delivered, placed, and finished above 85° F but below 100°F.

Winter months may require special cold weather requirements for placing concrete. These provisions restrict the contractor from placing concrete when the air temperature is 45°F and falling. In addition, the contractor is restricted from placing concrete until the air temperature is 40°F and rising. Other cold weather restrictions include; protecting the concrete from freezing until it reaches a minimum compressive strength of 1500 psi.

FUNDED RESEARCH

Current FDOT funded research includes:

- Maximum High Temperature for Mass Concrete - Determining the maximum allowable temperature for Mass Concrete structures. We currently require the contractor to maintain a temperature differential between the exterior and the interior of the structure of less than 35 °F within mass structures. This research will provide us with the additional protection of a maximum allowable temperature within a structure as well as the maximum temperature differential.
- High C_3S and SO_3 Contents - The intent is to investigate the use of higher values of C_3S in cements with our local materials. Additional research is being performed to determine the effects of high SO_3 (greater than 3.0%) in cements batched with local materials.
- Modulus of Elasticity (MOE) – The MOE of concrete mixes with local materials will be determined. The current values of MOE are based on concrete values that were developed approximately 25 years ago. The Department will study the changes to these values based on the current materials and batching techniques. In addition, creep and shrinkage will be added to the project in an additional phase.
- Determination of Voids in Post Tensioning Grout Ducts Using Non-Destructive Methods – This information will assist the Department and the

contractor in detecting voids in freshly grouted post tensioned ducts using non-destructive testing methods.

FUTURE GOALS OF THE DEPARTMENT

In the future, the Department plans to optimize the use of cement and cementitious materials for superior performance of concrete structures. Prescriptive methods of concrete design have been utilized for years with relatively good results. However, a new level of performance must be reached to guarantee a desired 100-year design life. This could be done by requiring performance based specifications and allowing contractors the latitude to develop concrete mix designs based on specific levels of performance. To accomplish this, test methods should be developed, or national standards be adopted and implemented, that allow measurement of this performance. In addition, non-destructive test results should be included as the acceptance criteria for concrete performance.

The use of non-destructive (ND) testing is another goal of FDOT that needs to be developed. A true performance specification should include, as the acceptance criteria, a method of non-destructive testing instead of the traditional testing of compressive strength cylinders. The concept will require that the FDOT break from traditional acceptance methods of cylinder compressive strengths. One challenge will be to educate the FDOT on these new test methods. Engineers have to understand the positive attributes and the shortcomings of ND tests so that they can make determinations and accurate evaluations of structures. The use of match cure cylinder tests and the use of maturity meters needs to be used to represent concrete curing and the strength of concrete under field conditions.

Due to the emergence of high strength concrete, the use of smaller size cylinders and their correlation to the standard size cylinders, needs to be determined. Also, there is a need for determining the correlation between flexural and compressive strength tests for local materials. In addition, research into the use of self-compacting concretes is needed. The effective use of these materials could assist in the placement of concrete and decrease the honeycombing that typically occurs as a result of congested steel areas in structural members. Proper proportioning of the mixes is needed to produce an increase in durability. The effective use of these materials could greatly benefit the Department and prestress/precast producers by providing a more uniform product with little or no bug holes and voids in the finished structure.

Recycled materials produced in Florida should be utilized to the fullest. There is no reason to import recycled materials like fly ash for use in concrete. FDOT currently has one source of petroleum coke fly ash and one source of bark ash as alternatives to coal-produced Class F fly ash. These waste materials are produced in Florida, so we have obligations to the citizens of Florida to use our own waste materials before trucking these materials in from other states. With that said, the use of recycled and waste materials generated in Florida is another area of research that is needed.

RECENT CHANGES TO OUR SPECIFICATIONS

The development of FDOT specifications is dynamic. These specifications are continually changing based on new developments in research and product evaluation. Some of the more significant recent changes are briefly mentioned in the following specifications:

Section 346 requires that the contractor now maintain the quality control of the project. That is, he is responsible for supplying qualified technicians for all materials testing on the project.

Section 921 provides new changes that require all cement producers to maintain a current quality control plan that identifies the issues of personnel, facilities, production, storage, and transportation.

Section 923 allows for the use of water from the washout of ready-mix trucks for the sprinkling of aggregate stockpiles and for the batching of non-structural concrete.

Section 929 requires that producers of pozzolanic materials and slag maintain a current quality control plan. In addition, it identifies the requirements for the approval of petroleum coke Class F fly ash, bark ash Class F fly ash, as well as bituminous coal produced Class F fly ash, slag, silica fume, and metakaolin.

NATIONAL INSPECTION ASSOCIATIONS

Based on the reduction of the work force, the Department intends to limit or eliminate inspections in the prestress/precast yards and batch plants. In lieu of the Department inspection of concrete and precast plants, a systemic improvement may be made by implementing quality control and quality assurance concepts and national certification institutes. Currently, the department is planning to require that plants participate in the national certification programs with the goal of quality improvement. Hopefully, these programs will offer the Department a higher level of quality based on the new qualifications.

CLOSING STATEMENTS

In closing, to meet the future needs of the Department for cement and cementitious materials, a joint effort between material producers and suppliers, academic research centers, and the FDOT need to be developed to ensure that beneficial research findings are implemented. Implementation of this research into specifications and test methods will ensure that concrete produced in the future meets the needs of the Department and the motoring public of Florida.

Photo 33: M. Bergin

Photo 34: K. Scrivener, J. Skalny and Å. Skarendahl

Photo 35: J. Tedesco

Photo 36: F. Jachimowicz (seated), D. Myers and S. Mindess

ENVIRONMENTAL CHALLENGES FACING THE CEMENT INDUSTRY

David Cahn
California Portland Cement Company
2025 East Financial Way
Glendora, CA, USA, 91741

ABSTRACT

Perhaps two of the more crucial issues facing those involved with Portland cement manufacturing are energy consumption and climate change (more frequently referred to as "global warming"). These issues are very much interrelated. U.S. manufacturers, woefully behind most of their European and Asian brethren with respect to applications of process and product technology, ironically may have near term "opportunities" (potentially mandated by international protocols) to make the more significant reductions in specific energy consumption and specific environmental emissions. For the long term, however, the greatest need worldwide will be for breakthrough process and product technologies that will reduce these specific factors by an order of magnitude.

Annex I country GHG emission commitments as percent of 1990 emissions

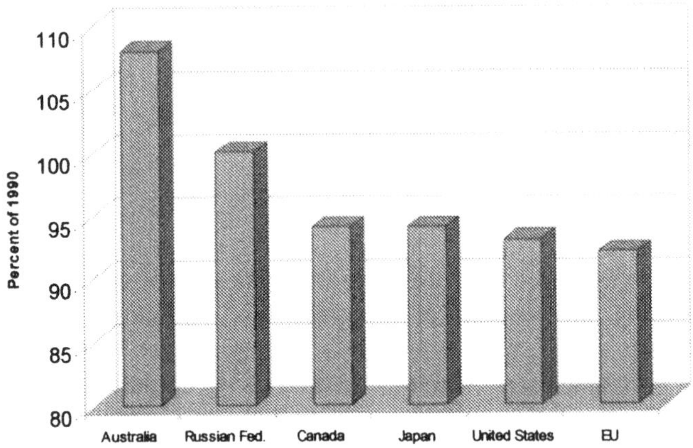

Annex I countries average annual growth in cement production 1990 – 1999

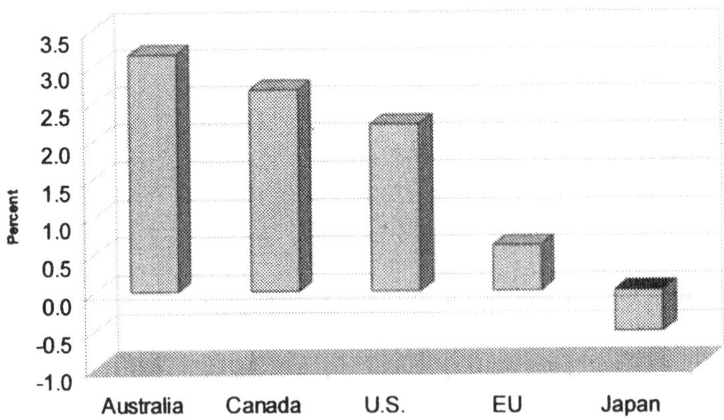

Cement production by country (million metric tons)

	1990	Avg. Annual Growth 1990-99	2012
Australia	6.1	3.2%	12
Canada	11.6	2.7%	18
U.S.	71.5	2.2%	114
EU	181.5	0.6%	200
Japan	87.0	-0.5%	76

GHG emission reduction opportunities

- Process improvements that focus on improved energy efficiency.
- Allocation of production to most efficient plants.
- Credit for use of wastes in the production process.
- Greater use of cementitious materials.
- Voluntary agreements with specific targets.
- Technology transfer with emission reduction credits.
- Participation in emission trading.

Credit for using wastes in cement production

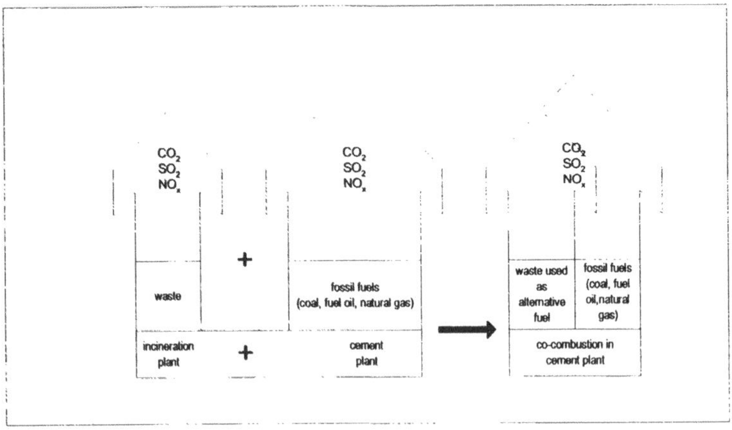

U.S. GHG reduction strategies

Cement production 1999	86 million metric tons*
Average energy efficiency	5.3 GJ per metric ton of cement
Estimated CO_2 emissions per metric ton of cement	0.9-1.1 metric tons
Total CO_2 emissions	79 million metric tons
Estimated average percent clinker per metric ton of cement	94%
Average annual growth 1990-1999	2.2%

Conclusions

U.S. has more room to maneuver than most Annex I countries

- relatively high energy consumption
- high clinker to cement ratio

GHG emissions per ton of product can be reduced

But strong growth in cement production may make it difficult to reduce total emissions by technical means alone

Photo 37: D. Cahn

Photo 38: C. Andrade and P. Trtik

UNIVERSITY-INDUSTRY RELATIONS REEXAMINED

Jacques Lukasik
Lafarge Corporation
61, rue des Belles Feuilles
Paris, France 75116

ABSTRACT

The first goal of any university or academic institution is to provide the highest level of education for its students. Research conducted in those institutions is essentially aimed at the progress of *Knowledge*, but it is also supposed to provide a ground for students' training and, in many cases, contribute to the progress of industrial competitiveness.

Companies, wherever they are situated and whatever sector they are operating in, are perfectly aware that the competitiveness of their products, processes and services (on which rely their own survival and evolution) largely depends on an intelligent and efficient exploitation of technology advancements and *Knowledge* in general. Always limited in their internal R&D resources, they have to count on scientific progress accomplished elsewhere, understand and evaluate it, and have the capacity to transfer and use it, in order to create a new economic value.

A proficient dialog between academia and industry is thus necessary. Because of cultural and language differences, such a dialog is not easy and often frustrating. This presentation will provide some thoughts related to this academia-industry interface and discuss how the communication between these two worlds could be improved, to the benefit of both.

LAFARGE

DIRECTION SCIENTIFIQUE

UNIVERSITY-INDUSTRY RELATIONS REEXAMINED

CEMENT AND CONCRETE : TRENDS AND CHALLENGES

Anna Maria Workshop II

Jacques Lukasik, Anna Maria Workshop II, November, 2001

LAFARGE

DIRECTION SCIENTIFIQUE

CONTEXT (1)

BASIC GOALS OF ALL ACADEMIC INSTITUTIONS

Progress of knowledge
Teaching and training

- **BASIC GOAL OF ANY INDUSTRIAL COMPANY**

Survival (staying ahead) in the competitiveness race
Creating value for its shareholders
Contributing to the social and economic progress of society

Jacques Lukasik, Anna Maria Workshop II, November, 2001

LAFARGE

DIRECTION SCIENTIFIQUE

CONTEXT (2)

- **INDUSTRIAL COMPETITIVENESS : INNOVATION IS THE KEY**
  - Fast evolution of technologies
  - Capacity to anticipate product and process evolution
  - Capacity to transfer and integrate
  - Efficient R&D is one of the answers...

- **INDUSTRIAL RESEARCH : FUNDAMENTAL SCIENTIFIC QUESTIONS**
  - Capacity to comprehend scientific progress
  - Limitation of internal R&D resources
  - Externalization of research

Jacques Lukasik, Anna Maria Workshop II, November, 2001

LAFARGE

DIRECTION SCIENTIFIQUE

...A NECESSITY FOR A PROFICIENT

AND CONSTRUCTIVE DIALOG...

...BUT THE COMMUNICATION IS OFTEN

DIFFICULT BY THE PROFOUND DIFFERENCES

BETWEEN THE «SCIENTIFIC» AND «INDUSTRIAL»

CULTURES...

Jacques Lukasik, Anna Maria Workshop II, November, 2001

LAFARGE

DIRECTION SCIENTIFIQUE

SCIENCE vs. TECHNOLOGY

- THERE IS NO TECHNOLOGY WITHOUT SCIENCE...
- SCIENCE STIMULATES TECHNOLOGICAL CREATION WHICH, IN RETURN, STIMULATES SCIENCE...
- ACCELERATION OF THE *TECHNOLOGY CIRCLE*

Jacques Lukasik, Anna Maria Workshop II, November, 2001

LAFARGE

DIRECTION SCIENTIFIQUE

BASIC SCIENTIFIC CULTURE

vs.

INDUSTRIAL, ENGINEERING TYPE CULTURE

Jacques Lukasik, Anna Maria Workshop II, November, 2001

LAFARGE — DIRECTION SCIENTIFIQUE

• **IMPORTANCE OF BASIC TECHNOLOGICAL RESEARCH**

To transfer fundamental research results : concepts, methods...
To create generic technologies and processes...
To develop innovative products which create value...

• **DIFFICULT TO ENSURE AN EFFICIENT INTERACTION BETWEEN THE THREE LAYERS WHICH CONSTITUTE THE TECHNOLOGY CIRCLE**

Science
Basic technological research
EVA Innovative products and processes

Jacques Lukasik, Anna Maria Workshop II, November, 2001

LAFARGE — DIRECTION SCIENTIFIQUE

BUT DON'T GET ME WRONG !

LET'S NOT MODIFY OUR RESPECTIVE ROLES !

• **Industry does not expect to have its direct product/process problems solved by academic research...**

• **Industry needs high quality University fundamental research, "UNDERSTANDING" of phenomena oriented...**

• **Industry does not need parametric type studies, in which the results apply only to a unique combination of systems and parameters and is difficult or impossible to generalize...**

Jacques Lukasik, Anna Maria Workshop II, November, 2001

LAFARGE DIRECTION SCIENTIFIQUE

HOW TO IMPROVE OUR COMMUNICATION AND COLLABORATION EFFICIENCY ?

- Increase the number of "translators" in the two worlds
- Increase University – Industry mobility, both ways
- Make industry needs better perceived
- Improve the "marketing" of research achievements
- Adopt a multidisciplinary approach to industrial research problems
- Favor comprehensive studies rather than parametric observations

Jacques Lukasik, Anna Maria Workshop II, November, 2001

LAFARGE DIRECTION SCIENTIFIQUE

Jacques Lukasik, Anna Maria Workshop II, November, 2001

LAFARGE

DIRECTION SCIENTIFIQUE

BRIDGE - SEOUL, KOREA

"a pre-cast segmental 6@20m, pre-stressed double tee"

Rc : ~200 MPa, Rf : ~45 MPa

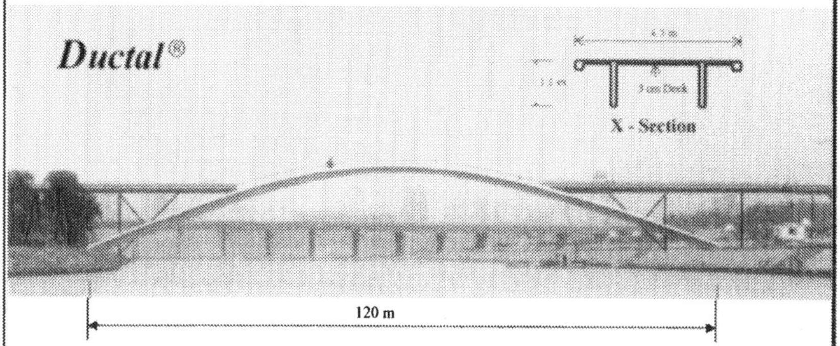

Jacques Lukasik, Anna Maria Workshop II, November, 2001

LAFARGE

DIRECTION SCIENTIFIQUE

Ductal® RESEARCH PROGRAM

- **A unique collaboration between three French industrial companies**
- **An intensive internal R&D program : about 40 people involved**
- **More than 10 university laboratories involved in a 3-year *multidisciplinary research program, related to basic properties and phenomena* concerning DUCTAL (2 doctoral thesis + contracts)**
- **A major coordination effort with excellent reporting**
- **Project managers with strong scientific and industrial backgrounds**

Jacques Lukasik, Anna Maria Workshop II, November, 2001

LAFARGE
DIRECTION SCIENTIFIQUE
Ductal®
Jacques Lukasik, Anna Maria Workshop II, November, 2001

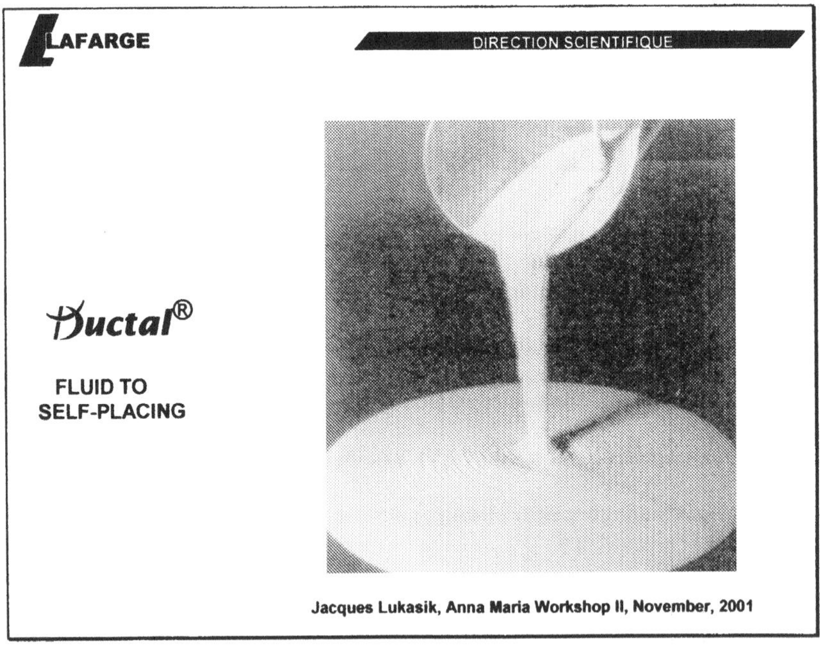
LAFARGE
DIRECTION SCIENTIFIQUE
Ductal®
FLUID TO
SELF-PLACING
Jacques Lukasik, Anna Maria Workshop II, November, 2001

LAFARGE DIRECTION SCIENTIFIQUE

Ductal® has the following characteristics:

- Excess fine particles at different scales (spacing factor),
- Very low w/c, typically half the stochiometric value,
- Low degree of cement and binder reaction,
- Low capillary porosity, typically less than 2%,
- Disconnection of capillary pores for low degree of reaction,
- Excellent transfer properties.

Jacques Lukasik, Anna Maria Workshop II, November, 2001

LAFARGE DIRECTION SCIENTIFIQUE

SOME OF THE EXPERIMENTAL TECHNIQUES USED TO STUDY AND DEVELOP *Ductal*®

- *Radiation-matter interactions* : XRD, SEM, TEM, NMR, IR and Raman Spectroscopy...
- *Thermal analysis* : DTG, DTA, controlled rate TA, DSC...
- *Interfaces and porosity* : water vapor and nitrogen sorption, Hg porosimetry, He pycnometry...
- *Mechanical and rheological methods* : nanoindentation, pull-out tests, shock and abrasion tests...
- *Durability methods* : electrical conductivity, accelerated tests...

Jacques Lukasik, Anna Maria Workshop II, November, 2001

LAFARGE

DIRECTION SCIENTIFIQUE

TODAY'S OTHER APPLICATIONS OF *Ductal*®

- Footbridges in Canada and in France
- Cooling towers in a nuclear plant in France
- Dome-shape roof in a cement silo in the US
- Acoustic panels in the Monaco railroad station
- Anchor plates for earth wall reinforcement on Reunion Island
- Safety vaults
- Manhole covers
- FHWA test program underway

Jacques Lukasik, Anna Maria Workshop II, November, 2001

LAFARGE

DIRECTION SCIENTIFIQUE

TECHNOLOGY CENTER IN L'ISLE D'ABEAU

Jacques Lukasik, Anna Maria Workshop II, November, 2001

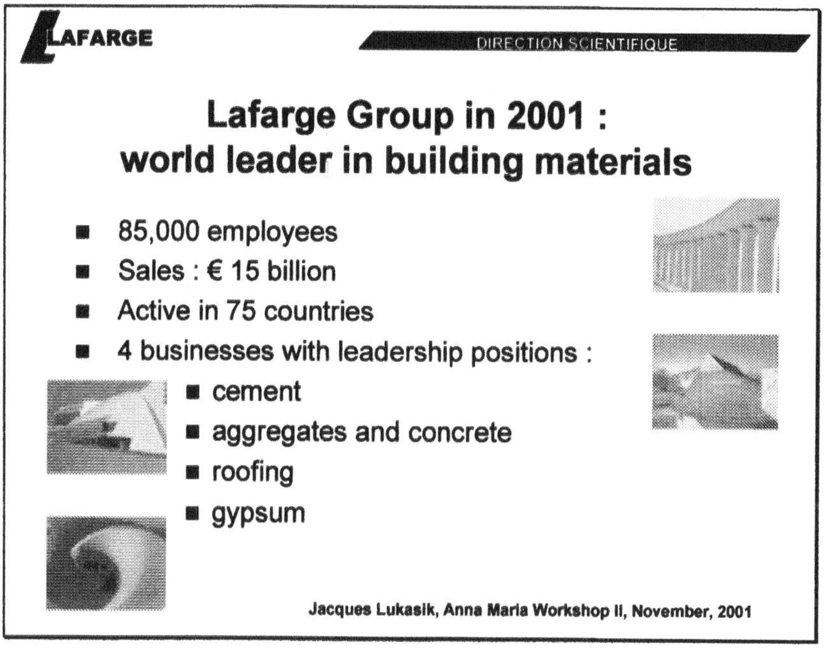
LAFARGE
DIRECTION SCIENTIFIQUE
Lafarge Group in 2001 :
world leader in building materials
85,000 employees
Sales : € 15 billion
Active in 75 countries
4 businesses with leadership positions :
cement
aggregates and concrete
roofing
gypsum
Jacques Lukasik, Anna Maria Workshop II, November, 2001

Photo 39: J. Lukasik

Photo 40: C. Drakos

NOTES ON R&D IN CEMENT AND CONCRETE

Juraj Gebauer
HOLCIM – Group Support
Holderbank, Switzerland
CH-5113

ABSTRACT

Considering the great number of patents, publications and international conferences on cement and concrete every year, one could assume that there is remarkable technical progress in our business. However, looking at the issues more deeply it becomes evident that in the past 50 years there have been few breakthrough inventions. Most of the patents are useless and many of the publications are just repetitions, with little or no contribution to technological progress.

The emphasis of the industry in the past was always on the improvement of the process and not so much on the product. Improvements were achieved mainly in efficiency of production and developing new production technologies. The existing, mature product - portland cement-based concrete - has many advantages that are difficult to beat. The problematic and expensive step from the laboratory scale to the industrial scale and market introduction, as well as the conservative standard specifications, make the introduction of novel products difficult.

A more focused activity of the R&D community is desired. The focus should be on:

(1) detailed basic understanding of the hardening and degradation mechanisms of concrete, in particular produced with "modern" concrete designs,
(2) development of reliable test methods for durability performance of concrete, in order to enable writing of performance specifications, thus opening the door for new product development and commercialisation,
(3) development of methodologies for transfer of knowledge from academia to industry, from research to engineering practice,
(4) focus on training and education of all parties involved in the construction industry,

(5) development of novel cements with better environmental compatibility and durability performance.

Additionally, professional and standardization organizations should focus primarily on improvements in the quality of their products (dissemination of the best knowledge in the form of literature and educational activities, formulation and development of test methods and standards), rather than on expanding their business activities.

R & D in Cement and Concrete

- ~ 2000 patents/year
- ~ 30 International conferences (~ 50 papers)/y
- ~ 40 International journals (~120 papers)/y

➡ PROGRESS?

- Adequate, satisfactory (fitting to the interest of the building industry/society)
- Not satisfactory (little progress in the last 50 years)

Holcim

Holcim Group Support 1

04 03 2002/JGe/gbo
R& D in Concrete ppt

Inventions of the Past 50 Years

- Air entrainment
- Superplasticizer
- Microfillers (e.g., SF)
- High Performance concrete
- Self compacting concrete

Other developments

- Improvement in durability understanding (ASR,DEF,F/T)
- Alkali activated cements
- Sulfoaluminate cements

Holcim

Holcim Group Support 2

04 03 2002/JGe/gbo
R& D in Concrete ppt

Barriers to New Developments

- Gap between academia - industry
- Gap between R&D activities - industrial practice
- Standards and market acceptance (conservative industry)

Holcim

Holcim Group Support 3

04 03 2002/JGe/gbo R& D in Concrete ppt

Gaps in R&D - Industrial Practice

- In the process of product development the most difficult and expensive step is industrial scale testing and commercialisation
- Unsatisfactory transfer of knowledge from research to field/market – know-how is available but not used enough in the practice
- Better training/education on the job site is required
- Focus of the industry in the past was always on the process rather than on the product
  - from shaft kiln to 5 stage preheater kiln
  - from ball mill to vertical mill
  - from coal to alternative fuels
  - from manual operation to automation

Portland cement chemistry has not changed substantially in the past 150 years!

Holcim

Holcim Group Support 4

04 03 2002/JGe/gbo R& D in Concrete ppt

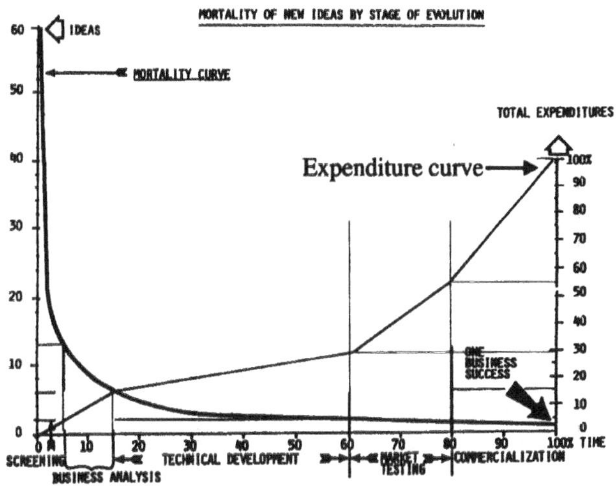

Holcim Group Support Name of Presentation/Event

Future Developments in Cement and Concrete

- No sophisticated and revolutionary developments can be expected due to:
  - large volume / low tech industry
  - conservative standards and liability issues
  - high risk and investment cost
- Any new development is evolutionary (slow) with very long introductory phase
- The "inventors" have no substantial competitive advantage over the "followers".

Holcim

Holcim Group Support 6

04 03 2002/JGe/gbo
R& D in Concrete ppt

Future Developments in Cement and Concrete

Portland cement is superior in:

- production cost
- performance
- raw materials availability
- environmental compatibility (?)

Replacement of PC is possible in:

- low grade applications (20-30% of the market)
- special cement (5-10% of the market)

Holcim

Holcim Group Support

04 03 2002/JGe/gbo
R& D in Concrete ppt

Research Needs of the Construction Industry - Conclusions

1. Better understanding of hardening and degradation mechanism of concrete
2. Reliable testing methods for durability performance of concrete
3. Knowledge transfer from academia to industry, from research to field/market
4. Better training and education of all parties involved in the construction industry
5. Development of cements with better environmental compatibility (low emissions, low energy consumption, recycling of by-products) and durability performance

Holcim

8

Holcim Group Support

04 03 2002/JGe/gbo
R& D in Concrete ppt

Photo 41: J. Gebauer

Photo 42: H. Zanni and J.-L. Zanni

Photo 43: C. Andrade and J. Lukasik

Photo 44: A.J. Boyd

CEMENT-BASED MATERIALS: A RESEARCH MANIFESTO

Felek Jachimowicz
W.R. Grace & Co.
62 Whittemore Ave.
Cambridge, MA, 02140

ABSTRACT

It is hardly an exaggeration when we say that our modern civilization is built on cement-based materials. The enormous economical and environmental importance of these materials in our daily lives, in every corner of the globe, is beyond dispute. At the same time, the lack of attention from the general population and public officials to this field is frustrating and discouraging to all those engaged in the area. Such a broad lack of interest leads to a lack of awareness of the enormous potential that could be realized in cement-based materials research and leaves this field bare of research funds. Improvements in performance characteristics, in environmental friendliness, and in broadening the scope of applications are areas where significant enhancements could be reached in the near future, if adequate funding was made available. A meeting among the many distinguished scientists in cement-based materials research is to serve as a springboard to create an ambitious and forward-looking document that shall become the blueprint for future studies in this area.

OPEN LETTER TO COLLEAGUES:

Dear Colleagues:

In our opinion, in spite of the enormous importance of cement-based materials on national and international economies, on the global environment, and in many aspects of our daily lives, this field remains a neglected area. There is a lack of attention not only from the general public, but also from government agencies. The funds assigned to this area are scant, and are not at all in proportion to the importance of cement based materials and structures in our society. The enormous economic, environmental and social impacts of these materials are not, by any measure, reflected in either the quantity or, regrettably, the quality of academic and industrial effort in this field. As well, the relatively small research investment in this area makes almost certain that the untapped potential of cementitious systems is unlikely to be realized in the near future. We believe that a strong and coordinated effort is required to reverse this trend.

With this in mind, we are turning to you for help. We hope that you and other distinguished colleagues and researchers around the world, in the areas not only of cement based materials, but also of colloid science, fracture mechanics, surface chemistry, self-assembling systems, biochemistry, theoretical chemistry and other disciplines, will join us in creating a ***Research Manifesto***. It is our vision that such a document could become a blueprint for cement and concrete research efforts worldwide, helping to:

- Facilitate the generation and distribution of badly needed research funds
- Establish ambitious, high quality and indeed *visionary* research and educational programs
- Attract excellent students and faculty
- Raise the general appeal of research into cement based materials
- Demonstrate the importance of cement based materials for the well being of our civilization

These are, of course, very challenging goals, which cannot be achieved easily or quickly. However, we hope that this proposal will not be viewed as being either naïve or presumptuous. We genuinely believe that, with your help and your insights, we do have the opportunity to accomplish these goals, which can have such far-reaching effects on the human condition.

Our concept is to create a manifesto with two integrated lists of research topics:

- The first list would address the *fundamental* chemical, physical and structural aspects of cement and concrete *science*. This list might have the heading: ***We wish to know.***
- The second list would address the *practical*, generally transparent benefits to concrete *technology* that could result from molecular level insights and the development of new methodologies. This list might be headed: ***Imagine that!***

We are counting on your **active** help and input into the creation of these two lists. We would also invite your critique and suggestions as to the format and structure of the Research Manifesto. Our plan is eventually to create a website that will serve as the main communication vehicle amongst all of the invited participants. This will enable coordination and integration of our collective efforts.

Please let us know as soon as possible if you would be willing and able to participate in this endeavor, by responding to the e-mail addresses given below. We will then keep you informed about further plans and progress.

With best regards, and looking forward to hearing from you soon,

Felek Jachimowicz (**Felek.Jachimowicz@grace.com**)

Sidney Mindess (**smindess@civil.ubc.ca**)

Jan Skalny (**jpskalny@aol.com**)

CEMENT-BASED MATERIALS: A RESEARCH MANIFESTO

Imagine that: we would have high quality, sophisticated academic research in the area of cement-based materials – visionary, fundamental research programs which would capture the imagination of bright young students, and inspire granting agencies and industry to provide more financial support for this area of materials science.
We wish: that this Manifesto becomes a blueprint that will serve as a guide and template for educators, policy makers, granting agencies, academics and industry for advancements in the field of cement-based materials.

Imagine that: we would have an adequate supply of highly trained experts in the fields of cement and concrete chemistry, physics and related sciences that could increase the level of sophistication in both academia and industry.
We wish: to establish university-based programs that will provide high quality training at both the undergraduate and postgraduate levels in cement and concrete science and technology.

Imagine that: we would have an in-depth understanding of the thermodynamics and kinetics of all of the chemical reactions and physical transformations involved in the cement hydration process.
We wish: to develop predictive models for both the microstructural and macrostructural behavior of cement-based materials, to guide our research and development into new materials in this area.

Imagine that: we would always be able to devise an optimal mix design using locally available raw materials for any set of performance requirements.
We wish: to develop algorithms for the optimal design of cement-based systems, using as input only the performance specifications and the characteristics of the raw materials.

Imagine that: we would be able to develop cement-based construction materials with vastly improved durability and service life characteristics.
We wish: to acquire a molecular level understanding of all of the issues governing concrete permeability and dimensional stability.

Imagine that: we would have a molecular level understanding of the phenomena responsible for the brittleness of concrete.
We wish: to develop technologies for increasing the toughness and reducing the brittleness of concrete.

Imagine that: we would no longer need steel reinforcement in concrete, thus eliminating steel corrosion problems and leading to lighter structures and a more rapid construction cycle.
We wish: to develop cement-based materials with sufficient strength and ductility so that steel reinforcement is no longer needed.

Imagine that: we would select construction methods and materials on the basis of the long-term needs of the public, and not primarily by the initial cost. Imagine further that road and bridge repairs would largely be eliminated, and the speed of new construction significantly increased.
We wish: to develop comprehensive, objective, easy-to-use economic models to permit the optimal selection of construction methods and materials. We also wish to enhance the durability of our highways and bridges, and to accelerate the speed of new construction.

Imagine that: cement-based materials would be viewed as being environmentally friendly, safe and attractive, and that both experts and the general public would see them as perfect choices for sustainable development.
We wish: to develop a clear and comprehensive understanding of the environmental impact of cement-based materials. This would include the development of materials that could be produced with minimum energy demand, materials that take advantage of undesirable by-products created by other industries, and methodologies for total recyclability.

Imagine that: we would ensure that the codes and specifications for cement-based materials are always followed, so that natural catastrophes no longer claim the lives of innocent victims because of shoddy construction practices.
We wish: to develop smart, self-policing materials that, when the codes and specifications are not met, will display a visible and unmistakable manifestation of this fact.

Imagine that: we would ensure that all cement-based materials always had a predictable performance, independent of the weather conditions at the time of placement, or the skill of the construction crew.
We wish: to develop simple test methods and instrumentation that will permit perfect quality control of the concrete during transportation and placement.

Photo 45: F. Jachimowicz

Photo 46: N. Thaulow and O.M. Jensen

DISCUSSION SESSION - WEDNESDAY, NOVEMBER 7, 2001

Mindess: Unlike some previous conferences that we've all attended, the intent here is to provide three quarters of an hour or so of some, I hope, fairly free flowing discussion about the issues that we've raised today. A number of us have spoken, one way or another, about some of the challenges that face the industry. Given the economic downturn which seems to be going on, certainly in North America, I think there will be additional challenges simply due to economic pressures for the next few years at least.

So, what we're hoping is that we will collectively try and identify the most pressing problems that the industry faces. Some of you may feel that there are no problems at all, and that's fair enough. We will begin, over the next couple or three days, to try and identify at least some avenues that we can take which will attempt to better the situation in terms of both research, which many of us are engaged in, but also in terms of simply developments within the industry to make it more efficient and to make it operate better. Much of the money that we spend on repair of our infrastructure is really money that should be spent on other things, if we'd only do it right the first time.

Suderman: Jim Kirkpatrick, in his very interesting presentation this morning, mentioned the importance of federal funding and support, and then several of the other presenters alluded to the importance of an industry-university liaison. I just wanted to make the point from a person who's got one toe in the boardroom and one toe in the university lab, that it's easier to get government funding if there's some show of industry support. That can be in the form of industry dollars, it can be in the form of in-kind contributions, or in some cases, just simply letters saying that this research that you're doing can be a value. Government funding agencies like that validation, the fact that the work the university is doing is going to have some basis in reality and is going to be used. So, it's not going to get you billions of research dollars or it's not going to get you thousands of bright intelligent young minds, but it's something that I think should be included in the manifesto: the importance of industry-university relationships and the role that industry has to play in increasing that level of research.

Diamond: Along this general context, we spent some time this morning discussing how to raise the profile of the cement and concrete fields in academic circles and how to attract or interact with important constituents of the entire construction industry. I had a couple of thoughts. First, with respect to raising the prestige of the field within the academic sector. There is one simple way of doing this: supplying endowed chairs. In any university, endowed chairs confer prestige, status, cachet. I don't know of any endowed chairs of cement chemistry. I don't know of any endowed chairs of concrete technology. Word perhaps to those of us who dispense funds.

With respect to people or groups within the field that we really need to educate and influence, it seems to me, architects are the people most influential in some aspects of the construction industry and who suffer almost uniformly from an absolute lack of knowledge of concrete technology. I think that if we could attempt to institute short courses in schools of architecture, this would go some way to solving a few of our problems. The other group of people who, it seems to me, we need to address seriously are people who go by the title of construction management or construction engineers, particularly academic departments where they train generations of field and construction engineers who deal with concrete every day, make decisions as to what specifications need to be invoked, how to comply with them, but again, who uniformly as a class know little or nothing about concrete.

Brandt: This morning we heard that most of the principal problems are solved. I would never agree with this opinion. It seems to me that there is a "bright future" for problems in this field. If we look at our modern cars, and we compare them with cars twenty years ago, which also performed very well, they're completely different. I would like to remind you that in the late 50's a high level U.S. functionary in technology, I will not give his name, declared that there is no future in the computer industry.

Mindess: Thank you, and I guess that there's probably a danger in trying to manage these things too closely. After all, one of the products of American industry that was so closely managed as to be sort of perfect was the Edsel.

Jachimowicz: After this morning's session, I took a long walk on the beach trying to digest the essence of our discussions. I want to share with you probably a little more provocative note following on Andrew's {Brandt} statement. A couple of years ago a book was written. The title of the book was *The End of History*. About two years ago, I think, another book came out; *The End of Science*. Neither history nor science has ended. The end of science has been proclaimed for many

many years, many many times, but if you would have asked anybody around if we would have wireless telephones today probably nobody would believe it. I don't think we have solved all the problems in concrete. Yes, we can make perfect concrete today, absolutely. If we pay attention to it, we can.

But let me start from a slightly different angle. Look how computers are being designed. The designers of computers and software don't teach us how to use them. They don't come to us and say, "you guys have to learn how to use computers, you guys have to learn how to use the software". They ask a completely different question: "How do I design this piece of hardware, how do I design this piece of software so that it is absolutely easy to use, so that you open it and you know how to do it? Anybody knows how to do it. You don't have to have a Ph.D., you can be anybody and you can use it." I think the question for us is not necessarily that the training component should be neglected, the education component should be neglected ... absolutely not. But to rely on this only, I think, is naive. We'll never get there. I think we should, as people who design materials, we should ask ourselves a question. How do we design this material so that it is abuse-friendly, no matter what kind of people are going to use it or how they are going to use it, so that they are not going to misuse it or abuse it, and it will still work as designed? That's a very difficult issue. Of course it's very difficult, but I think it's a challenge, a challenge for anybody in this field.

Cahn: I would like to interject a sort of political, economic, somewhat cynical, comment but building on what Jim Kirkpatrick was talking about this morning. The means by which we can build political and general public support for our industry really involves funds at their basis. Most of these funds really derive from the cement production side of the industry and are directed by those in the so-called "corner offices" that run them. The cement industry seems to be fairly willing to devote rather large, comparatively large, amounts of money toward very short term efforts in this area. That is political advocacy that will get the next highway or airport approach appropriations bill to provide demand for the product or promotional efforts that are even public oriented to build concrete houses or things of that nature.

There seems to be less of a willingness to make a long term investment in the industry, which would be what we are talking about today; to build the research infrastructure, to bring in the capable people, to generate a continuing stream of funds from the government, particularly federal funding. That which is being devoted is usually the first thing to be thrown overboard when the economy turns sour, which it is doing now, and which is being done now, where budgets are all being cut back in the areas I just mentioned. So, whether it's a decrease in demand for the product brought about by general economic constraints and problems, or whether it is importation of cement by entities that have no concern about the long term viability of the industry and, as we had a few years ago with a very robust demand, but to put it bluntly, lousy pricing because of imports, these pose very significant problems towards creating long term development of the kinds of advocacy that we're talking about. Dr. Shah, I think, is well aware of this problem with his group that he is trying to promote and pursue in that area, which is now very subject to the vagaries of decisions by the cement people as to "do we or don't we?" Where do we put our dollars? Long term to promote an industry or short term to promote a ton of sales next year? That's the problem.

Kirkpatrick: Since I've been tarred with this first order scientific question business, let me restate my point there. My point was not that there are not first order questions. My point is that it is the perception of university administrators, it is the perception of people at the National Science Foundation. Those kinds of people believe the first order problems have been solved. My point is that the need is to communicate the vitality of this industry and the vitality of the science and technology to those people so that we can develop a research industry.

Roumain: I think that there are a lot of needs for more research, but the biggest problem that we face is that there is an abundance of existing technology that is not finding its way into the marketplace. I feel that one of the biggest problems is, and we heard it this morning from Dr. Andrade, from Jim Kirkpatrick, and from a lot of the other presenters, that we have a fragmented industry and there is no one voice that speaks for all of us. One perfect example, sponsored by the Department of Energy, is Vision 2030. Here we have another set of brains that are in parallel but not communicating with each other. The left brain is not talking to the right brain. If you can put those two parts of the brain together I think we could be well on our way to being one of the industries of the future.

Mindess: That's an interesting thought. Is there some way to get the industry, at some level, under a single umbrella. Mostly, they sort of fight with each other.

Trtik: As I look around, the reason why I took the microphone is, I think, that I am the youngest one here. In the morning there was quite a discussion about how to attract young people and when Sidney Diamond talked about why this industry doesn't have endowed chairs, well maybe it's because it doesn't have students. It's a vicious circle. If you get students, you'll get endowed chairs. I don't know but that's my feeling. The trouble is; why am I the young guy here and I'm sorry, you are all great, but I don't see many more people of my age.

I must also recall the words of Sid Mindess regarding the industry which we are talking about. We can see it everywhere around us and what you see is people, it was originally 30 years ago, who can speak four words of English, two of them are "more water". I can see the same thing in Prague. This industry is run using unskilled people, often illegal immigrants. Is that sexy for young people? I don't think so. Isn't that a little bit the mistake of industry that it doesn't make it more prestigious? One idea which sparked in my head during this discussion is how about putting a little bit more accent on precast concrete. I know that it has got many limitations but you can always take precast concrete, this segment, and make it beautiful. Cladding of one of the pivotal examples of architecture in Prague of the late 90's is *Dancing House* on the embankment. It is made of concrete, precast concrete, with a precision of less than one millimeter. These are huge blocks, three by three meters, four by four meters, something like that, and people do not know about it. I'm just wondering why.

I got my Ph.D. a year ago and when I went back to Prague, people asked me questions. *So, what did you do in Scotland?* I said, well I got my Ph.D. *In what?* In concrete. *Wow, in concrete? Are you kidding me? It's sand, water, cement, we know all about it. How the hell can you get a Ph.D. out of it?* Well, maybe that's a little bit a mistake of academia. It doesn't provide good publicity.

Coverdale: I guess I have a question here for the group. I'd like to see a show of hands of people who represent concrete producers or specifiers. Three? The group I think we may be missing here is contractors who place concrete. A lot of the quality issues that we're talking about are in the hands of those groups as opposed to in our hands. I guess one comment is; that's a big segment of the industry who's voice isn't here and some of the cement producers and perhaps the admixture producers are trying to interpret or maybe represent industry in that sense but I don't know that we can always do a good enough job of representing their needs in a situation like this.

That gets me to my next comment, which is, we've talked about a lot of other industries, the auto industry, the chip industry, computers and how they're different. I guess one of the things that seems clear to me is that there's no economic incentive for the producers, in our current financial model, for producers and contractors to make durable concrete like there is for a chip maker. It's easy to put a lot of money and a lot of investment, there's a lot of dollars that can be had for someone who comes out with the next computer chip, but I'm not sure that there's a lot of money for the person who can make something last 60 years versus 50 years. And so, I don't know that we're dealing with the same economic models that drive research and that drive technology development. That may be a fundamental difference between our industry and some others. Maybe creating awareness is a way to help that, but again I think it's an issue of shortsightedness versus longsighted because the kind of value we're talking about bringing to concrete is long term value and with most of the other industry comparisons we've made it's short term value. So, I think we might have some uphill battles to convince people of long term value issues.

Mindess: Are you suggesting it would be in the industry's best interests to make increasingly less durable concrete, sort of an enforced obsolescence that no shortage of industries do use. The toy industry, for instance, expects a half-life of about 20 minutes.

Boyd: Actually, I would like to address the point about no contractors being here. I actually did contact a number of them to see if they would be interested in participating, from the end user aspect. The consensus was that they didn't want to get involved in anything that would make their lives more difficult, which is what they seem to consider progress to be.

Lukasik: The durability aspect, I think that this is our thought in our industry. I mean we are faulty because we are not able to market the usage value of our products. We've been too long in the commodity sort of business. Now, it's becoming a sophisticated product. Cement is becoming sophisticated. We have different types of cement. Concrete is becoming very sophisticated. It has different usage values that we are not able to sell because we don't have marketing people who are smart enough, maybe. We have not gone this way but at Lafarge we are picking marketing up and we are trying to connect marketing and usage values for our customers. We try re-pricing different values of our products that we want to sell but it will be a long way because you do not turn it around like this. That's the way I feel. Durability, one day, will be valued, will have a price and you will be charged for it, sort of a guarantee.

Hooton: Just a small comment. You mentioned about the contractors. The current president of ACI is a contractor, Baker Concrete, I think it's Don Baker who's the president. Anyway, one of his initiatives this year is to set up a whole series of scholarships. He's trying to raise in the order of seven million dollars to endow scholarships for students, to attract students to the concrete industry. The first three will be announced soon, there were six finalists, and he's making this a commitment before his term's out to raise this money from the whole industry, to push for bright new students in concrete, and this is a contractor doing this.

Scrivener: I have two comments. One is on the role of research. I think very much that the leadership has to come from the industry. It's all very well talking about getting federal funding but most companies working in the sector are investing far below one percent of their profits in research, which is a pitiful fraction of what you get in other industries. Well, that situation continues and the companies show that they don't think research can sell anything, then we're not going to be able to pull everything else up. I think there are some positive moves on that front, there are some companies which are starting to invest more in research and in the long term. Hopefully, they'll show that they can get a payback from that.

The second comment, on the contractors. We organized the seminar in Lausanne a month ago and we did actually manage to get, not a contractor, but one of the major consultancy companies to contractors to contribute a paper on research needs. One of the points he came out with, which I thought was very interesting, is they're being confronted more and more now with having to make assessments of structures as they are going back into the private sector, so assessments of roads, assessments of tunnel structures and things like that. Ironically, the fact that a lot of these structures are being taken over by the private sector who then have an interest in their durability, I think, can be quite a positive force. It seems that while things are built for the State, nobody's going to be around to care whether they've cracked up in five year's time. If these have to be managed by the private sector, then they're going to perhaps pay more attention to the quality with which they're built in the first place.

Diamond: With respect to the potential market value of durability, the thing that troubles me is the concept that there is such a thing, such a single thing, as durability of concrete. We have a bunch of different, identifiably different, syndromes. Some concretes have steel corrosion problems. Some concretes have alkali-aggregate reaction problems. Sometimes you can change things so that all of the potential problems are benefited simultaneously but often that's not the case. Increase, for example, the alkali content of cement for a particular application, you may be benefiting the potential for steel corrosion, but obviously reducing the potential durability if you have potentially reactive aggregates present. So, the idea of a single entity of concrete durability seems to me a rather extreme simplification that we need to get beyond. In human terms, it's analogous to, I suppose, the average life span. We bring up the average life span by curing diseases one after another and the net result is durability but there is no one thing we can do to people to uniformly increase their life span.

Lukasik: Maybe I was a little bit misunderstood. What I was saying is, you mentioned ASR or corrosion, we know how to handle, how to prevent that. We know how to make concrete that will not exhibit ASR problems. If a company that sells concrete shows, or guarantees, that this will not exhibit that, takes the responsibility for it, then it can price its product differently. So far, nobody is doing that and probably it will take some time before somebody does do it but the car industry guarantees 100,000 km without corrosion of the body and there are other examples where problems are guaranteed. So, I'm not saying that this is the immediate thing but somehow we will get to a better, probably a higher, price because if you guarantee something you take responsibility. I think that over some years in the future we will, and durability is only one example of what I call the usage value for our customer. There are some other values that could be guaranteed as well.

Mindess: You could take a lesson from the banking industry, who find a new service charge for absolutely everything they do these days.

Andrade: About the issue of durability, I agree that it should be paid for! I mean, the 50 years of expected durability comes from the structural engineer. They are thinking about the period of repetition of earthquakes or extraordinary loads, but from experience we know that in many environments concrete is not durable for more than twenty or thirty years. In that case, if we say that the concrete is going to last, and it is not lasting, this is why they don't trust us. I think that if you'd like to have fifty years of durability for concrete in a harbor, for instance, I would tell the authority of the harbor "you must pay for the higher cost of this concrete, you cannot have it for the same price as normal concrete". In that case, I agree, that it has to be paid for and this has to be taken into account. This is one of my comments that I tried to explain to Sidney.

The other point I would like to mention is the following. I was talking to cement producers and to constructors. At least in Spain, and in many other countries, they are doing very well and they are earning money and they say "Why fund research?" We have very high benefits, and to produce more benefits, research will not make a big difference. In that case, the crisis is not in the industry. The constructors say, "We assemble components, we learn how to assemble components, we manage the construction, and we install whatever the component is. We are able to build and it is not a problem. We don't need research." In that case, I realize that I need to address the owners. The owners are those who are really concerned and I think that we should address the owners. When they are public owners and they are politicians, they change frequently, so you cannot completely trust them because their policy may change. The new trend, as Karen

{Scrivener} mentioned, is to have private management of public works. This will make a big difference because when you have a big owner like electricity generation or even the people of radioactive waste management, they need to keep structures in good shape for a long term. In that case, we should address the owners, and the permanent owners, of the structures if we want to get funding for this kind of sector.

Shah: I'm going to comment on what a couple of people said. A couple of people mentioned, and although you probably did not imply it, you mentioned that we have enough technologies. The question is transferring them. I've heard these comments, especially when you go to ACI meetings, for instance, "Well, we really don't need new research if we'd just use what we know and transfer it." It may be true as far as existing short-term problems but if you really want to think *out of the box* then we really need to have knowledge of its history. For one example, can we not monitor the concrete right from the beginning just as we are monitoring human health in prenatal children?

I mean, it's really possible that if we can put our emphasis so that immediately after it is cast we can develop sensing and monitoring tools so that the construction worker doesn't have to be knowledgeable, he just looks at the black box and says, "Ah, it's red, something is wrong." For this kind of thing, I think we really have to combine the knowledge of materials science, sensing technologies and various other disciplines. This is only one example. This is the kind of thing where we do need industry and government involvement so that we can get the rest of the nation excited. We have to be careful when we say the real lack is technology transfer and we don't need any more research.

Skalny: I would just like to comment on what Carmen {Andrade} said. Many years ago, we had a committee on concrete durability at the Academy of Sciences. This was in the early 80s and one of the committee's recommendations was that, yes, the responsibility should be with the owner because he is the only one who can force the suppliers, and the others involved, to do their job. So what happened? With great difficulties, we published about 600 copies of this report. The Academy of Sciences had no money to publish it, so I had to convince Martin Marietta Corporation, my employer at the time, to finance the publication. When we finally distributed it, there was not a single response other than one or two criticizing us for even attempting to change the ways industry works. I think we know what would be correct, but the issue is how to convince somebody to do something about it. That's where the problem is.

Kirkpatrick: This is more a question, I guess. Folks that I've talked to in the industry made the comment that when you run an economic model, you use a discount rate of five percent or so, durability counts for nothing. And I guess I'd like to see what the opinion of this group, at least, is in terms of how the industry runs its economic models and what aspects of those economic models are controlling the marketplace.

Forster: I guess I'm one of the few, perhaps the only, federal representative here and the question I keep hearing recurring here is how do we get the industry interested in doing research. From the owners perspective or the user's perspective, to this point our specifications have been basically prescriptive in nature so with that type of spec, there is really no onus on the contractor. The onus is still on the owner because you tell him what to build. So, from our perspective in Federal Highways, we're trying to transition from prescriptive specifications to QA/QC specifications to performance based specifications so you tell the contractor what you want, in terms of performance, and he has to produce it. Kind of following on from that will be the warranty situation, which I think we've heard several people refer to, in which the contractor has some ownership and responsibility for what he builds. Then, all of a sudden, if he has to provide performance and he has to warranty his product for a certain period of time, then naturally he wants to make it better and cheaper at the same time, so perhaps research can help him to do this.

Diamond: Jim {Kirkpatrick} raised the question of economic models and Jan {Skalny} may remember, that in the NAS group we had some discussions on this. We were told that the structure of the economic model is driven primarily by the depreciation rate and it so occurred that the laws governing depreciation at the time, in the United States at least, were such that it was uneconomic for a private owner to own the structure more than five, or maybe seven or eight years, that he has to sell the structure and get the depreciation cost going again, which means that the owner that builds a structure has only a very limited interest in how long that structure survives. If he can turn it around, sell it to somebody else after a relatively short period of time, and if it isn't obviously crumbling, he's out of it. Somehow, the economics of that escapes me but that seems to have been the situation. Whether it is currently so or not, I don't know.

Roumain: I personally believe in all types of research, from basic research to applied research to part of development, whichever name you want to attach to it. We must understand that we must see some form of light at the end of the tunnel and when we are talking about developing, whether it is a new technology or a new concept or a new service, there are some common denominators that must exist. One is, is it making our process more cost effective or is it bringing more to the table that we can afford to pay. That's extremely important. Second, we must make sure that durability comes into part of the life-cycle analysis that must be weighed. Also, environmental impact is the new form of bringing technology to market. Are we becoming more environmentally friendly with the concept we are putting forth. We must weigh all these different values when we say we are going to bring something new to be accepted into the field.

Gebauer: I would like to comment on the economic model that Sid {Diamond} and others commented on. The cement industry is very much interested in economic models. I mean, this is all about industry. They are interested in making money. In the present situation, I think, it is very difficult to put durability into the model. You can ask any salesman. It's much easier to sell early strength because you can convince your customer that he will get better early strength, but it is difficult to sell durability at the present state of specifications and codes of practice with the situation that we have in standardization. So, I think in this area we must change something in order to really value durability and other the properties that Jacques {Lukasik} mentioned in his contribution. I think this is where we have to do something about how to define durability, how to test durability, in order to have a clear definition and appropriate testing methods. These are still missing so we still have lots of things to do in this area.

Ma: I agree with Dr. Shah's comment that it is very important to transfer the existing technology to the industries but it's also important for researchers to understand industry's interests and needs. As a cement producer, we are no different than any manufacturing industry. Every day, we worry about two things, production and cost. You know, how much product we can make every day, how much does it cost to make those products and what is the cost of service to the customer. So those are two things that are no different from any manufacturing industry. But, the difference is, we spend much less money on research and development. In the cement industry in general, we spend an estimated 1% of sales on R&D. The problem there, I can see, is that I often receive proposals from universities and, in most of the cases I agree with California's cement industry {CPCA} explanation, their intent to focus on short term returns. In most of the proposals we really can't see those projects that can help us to sustain competitive advantages and we can't see any short term return. So, basically, we turn down a lot of proposals. I think my point here is the communication between researchers and industry is very important and the researchers have to understand what we want and work with industry to generate some project which is mutually beneficial to both parties.

Andrade: I would like to mention something that has not been discussed. In RILEM we had a meeting of the Board of Directors, and the Japanese delegate presented the priorities in Japan. It looked like the civil engineering field in Japan is still an important area and I realized that safety, the safety of people due to earthquakes and disasters, is a big concern there. Of course, this is due to the climate and so on, but it reminded me that, in the case of my institute, we raised our social profile very much last year because two collapses occurred with some fatalities. We were asked to assess what happened. In that case, I realized that safety is one of the main problems for society. Safety really is very important. The coefficients of the factors of safety that we use are very high because we don't want to have any fatalities. But if you are able to communicate that safety, you are able to improve things about safety in disasters. I'm thinking about the twin towers attack and how they collapsed. We are able to deal with this subject in a much more modern way as risk analysis. That could also raise our social profile because we can save money and increase the feeling of the people about safety and safety related to the building of public works.

DISCUSSION SESSION - FRIDAY, NOVEMBER 9, 2001

Mindess: Ladies and gentlemen, we've had two and a half days of very interesting discussion focused largely around the theme of research, both in industry and in academia. We've seen discussions ranging from the very practical requirements of say the Florida DOT in terms of the concrete that they're making these days, everything from that to the nanotechnology that Pavel {Trtik} spoke about yesterday with such enthusiasm. I'm not sure how one bridges these things easily.

We've had two very interesting and remarkably similar presentations, this morning by Jacques {Lukasik} and on the first day by Felek {Jachimowicz} talking about the industrial view of research and that those of us in universities and in research labs should be doing truly fundamental research and leave the product development to them. It's hard to disagree with that. I assume that, implied in there somewhere, was their willingness to support fundamental research at the universities.

One of the things that I think we have to bear in mind as I listen to these discussions is that; many years ago C.T. Snow wrote a book about two cultures, contrasting the science based culture and the humanities based culture, if you will, and the great gap between them. I think we have a similar gap in the two cultures represented here between the industrial culture and the academic culture. How to bridge those is, I think, the problem we're in part going to discuss for the next little while.

I'd just like to make sort of parenthetic comment to complicate things. At least in North America, for historical reasons, almost all of the cement and concrete research is carried out in departments of civil engineering. Not all of the people are civil engineers, some come from other backgrounds but, nonetheless, that is almost uniformly where we are housed. It's hard to think of a more inappropriate place for us in terms of trying to do fundamental research, or to train people to do fundamental research, because inevitably we are dominated by the structural side of the department and there's just very little support in the end for fundamental research of the type that we've been talking about. So, one of the things you might want to think about is where academic cement and concrete research should be housed because we don't serve you very well by and large in the civil engineering department, no disrespect to the civil engineering professors in this room.

I think that the other theme that came out, and perhaps that's where we should start, is the general view that somehow cement and concrete research has to be given more prominence. Whether it's in the universities, whether it's with governments, whether it's the funding agencies, we have to make a much stronger case for the importance of our industry. You've seen some of the numbers in terms of the magnitude of what collectively we produce and this seems not to be recognized by the world at large. So, with those few remarks, if I can just get the discussion going, what we would like to do by the end of the morning, if we can, is have some kind of rough road map or guide as to how we might improve both the quality of research, the direction of research, and perhaps ways of getting it to be a little closer to the forefront of the minds of our governments at various levels who end up funding so much of the research.

Marchand: From someone who's actually teaching in a civil engineering department, I guess that the two presentations made by Jacques Lukasik and Felek {Jachimowicz} were pretty interesting. I guess that industrial partners should state more clearly that they want more basic research. Coming from a civil engineering department, I can tell you that for every industrial partner asking for more basic research, you've got twenty others asking for applied research and also for the 'D' part of the R&D process.

Out of these twenty industrial partners, you have at least ten of them who want you not to limit yourself to the R&D parts, but to implement the technology within the market. These are usually smaller industrial partners and they've been extremely vocal over the past few years. This has had important consequences with respect to research. Those of us who are coming from Canada know that you cannot apply for research money without an industrial partner now. That's an important factor. Not only do you need to have an industrial partner but you need to justify the impact of your development work on the actual market to get the money. That's one consequence.

The second consequence is that, within civil engineering departments, materials research is considered to be a low-tech type of research so there's less money available for civil engineering. The National Research Foundation in Canada has substantially decreased the amount of money given to civil engineering research and, out of this money, there's less money available for research on materials.

The third consequence is that, since you have less money, you have fewer teams working on these topics. Well, it's certainly the case in North America. If you look at the number of teams that were working on cement and concrete research 20 years ago and now, there's a huge difference. So, you have fewer students. The number of students learning about concrete has been reduced so you have less people likely to join industry and increase the profile of our industry.

For the last point, the last consequence of all this, I want to take the example of modeling that was raised by Sid {Diamond} yesterday. We have good examples of models that haven't been validated. Well, in many cases, these models have been funded by industrial partners who are asking for a very simple model and they are requesting those who are doing the actual research to make the model public as quickly as possible. They don't care about validation. So, who should we blame? The person who is doing the research or the person who is pushing the researcher to make his research public as quickly as possible without any validation?

I mean, there is a real danger here. If we're not going to have enough people working on these topics in future years, the industry won't have any safety net and the quality of the people involved in our industry is going to decrease dramatically. I guess the point that I'm trying to make is; we need more industrial partners stating clearly that they want good quality basic research. That's point number one. Point number two, people from university should clearly understand that if we're not doing high quality research, we're increasing our own problem.

Mindess: What you're saying is that industry should divert money from its lawyers to its researchers.

Roumain: I'd like to deviate a little bit because I keep hearing basic research versus applied research. There is one other set of research that I think is going to be resulting from the meeting today. It's something we've been doing over the years. It's what we call defensive research. Yesterday, and the day before, I heard three comments on thaumasite, from Sid Diamond, from Norah Crammond and from Doug Hooton.

The reason I'm making this comment is, as a product manager for my company, my job from the product side is to help create value to our shareholders, just like Jacques {Lukasik} mentioned this morning. To keep creating that value, we must satisfy the need of our ultimate customer that is buying and has to live with the ultimate product. Now, when I hear the word thaumasite, it gets me a little bit nervous because I'll be the one going to ASTM at the next meeting to promote the use of up to five percent limestone addition to be incorporated into ASTM C150. Am I on a collision course with the three experts that talked yesterday, and am I going to have to create more research to defend myself out there.

Before I hand over the microphone, I would like to quote what my colleague Lesley Ko mentioned. "If you choke once on your food, do you stop eating?" Is that the Chinese saying? You may want to quote it better than I do Lesley. So, I would like some answers on that because I'm definitely going to see lots of papers written about thaumasite and then we're going to have to go back to the universities and say, do this research and do that research, to defend our decision that limestone addition is a good way to go.

Hooton: Well, just because you raised it. When we first did our work on thaumasite, it came out of the blue. We weren't expecting it. We had a severe deterioration problem. When we went to publish the paper on it, the cement industry, largely the European cement industry, tried to prevent the publication from being made. They said it was thermodynamically impossible for this to occur and they truly tried to stop the paper from being published. I know Norah Crammond did work for years and the industry wouldn't let her publish papers on it because they didn't want to hear the answers or hear the question.

I think the question is, I've never seen any damage from five percent carbonate addition. We do it in Canada. There are cases that I have been involved in with thaumasite that don't involve carbonate addition cements. Type V cements are now allowed to have carbonate additions in Canada and the cases I've been involved with are Type V cements so I'm not so sure. I just read a paper yesterday, just at lunch, and it just happened to be by Colaparti. He showed some data indicating that ten percent carbonate additions did cause thaumasite damage in the laboratory with relatively modest sulfate exposure. So ten percent, in at least one paper, showed that. I think it is something that needs to be looked at but not addressing the issue doesn't make it go away.

I'd like to raise something else that Sid {Diamond} and I were just talking about. Nobody's mentioned the word Pyrament here. Ten years ago, we heard; "Pyrament, this is going to be the best thing since sliced bread." We went down to the TRB meeting and the president walked in with all his people behind him and he told us how he was going to save us from everything. This thing came and it peaked and it died. I know that there was a lot of interest in what it could do but my perception of what happened is that they didn't do all their homework. In Toronto, they tried three jobs at the airport and every time a truck arrived it was different concrete. One time it would set in the truck. The next time it was fine.

The third time it didn't set for two days. It was just non-controlled because they hadn't done their homework. Maybe there were other issues but it drove a large company into bankruptcy.

Similarly, nobody's mentioned thaumasite. We've mentioned the success story but nobody's mentioned this failure, the large failure of a technology. I would just like to mention, to echo what Jacques {Lukasik} said about not wanting parametric studies from universities. Your own company in Canada is asking us for parametric studies and they think it's going to fund a Ph.D. for them, to get a Ph.D., but it doesn't because you can't get a Ph.D. from just some testing and analysis. We have to go into government funds to get the mechanistic work done that will actually get a Ph.D. for that work. The parametric work and the product development work, which they're also asking us to do and which is part of the project, is what they're willing to fund, not the basic research. So, I think that the reality may be different than the perception.

Lukasik: I would just like to correct, maybe, a misunderstanding. I didn't say that we had, in the industry, something against parametric studies. Very often you need to go through the step of parametric studies before you understand what is going on. But, we see so many papers published on pure parametric studies coming from the university community today that I'm criticizing this point, in particular. If there is a parametric study because, behind that as a final result there will be an understanding of the mechanism, it's fine.

Now, the second point is that if the industry asks for this type of study, it is probably the university researcher who should correct this statement of the problem. So, that's what I wanted to mention. As far as Pyrament is concerned, the problem with Pyrament is that, when it came out, Lafarge had a similar product within six months, except that it was durable. If I remember correctly, around 1988 when Lonestar came out with Pyrament, the main problem was durability. It wouldn't pass some ASTM standard durability test, I don't remember the number of the test but that was the main problem. Hot water, for example, with some chlorides in it would make it totally mushy and fall apart.

Pierce: You've already heard from the other two owner reps that are here, Steve Forster from Federal Highways and Mike {Bergin} from the Florida DOT. I represent the Bureau of Reclamation, which in this part of the country is little known. We only work in the western U.S. It is true, in fact, that we have made contributions over the years to concrete technology but from our perspective, now we are no different than industry, than academia. We have to fight to get research funding and the way we see out of that hole is to do joint venturing. That gains us a lot of credit when we can go to our administrators and say, this is an industry need, it has applications for us in our work, and we can joint venture. We have facilities. Juri {Gebauer} mentioned this morning about having larger testing facilities and typically in academia you do micro-size research. We have facilities that include, for example to those of you who have not seen our facilities, a five million pound universal testing machine. We have large facilities for doing things. Unfortunately, I admit, those facilities are now largely idle because of our problems.

Jim Kirkpatrick mentioned that those first order problems seem solved and that's the administrator's perspective. We spend more money now trying to figure out how to get salmon upstream for spawning, which is important, but if we are to have long term durable structures we need more understanding of how and why. It's very difficult to get our administrators to understand that. In fact, back in the 70s we had a dam that was suffering from ASR and our solution was to tear it down. We ended up rebuilding it using a very low alkali cement with a natural pozzolan, belt and suspenders to nth degree.

Our more recent problems is the Durret Mountain dam in Arizona. All we did was say; okay, we have a problem and we just post-tensioned the structure in place. So, we have a damn that has exhibited ASR over a number of years and we didn't do any research to try and help that issue. So, we've got to do joint venturing. Our biggest research need is durability. Certainly, there's a strong push for performance specifications, but there we sit spending the public's money and we need to grant more and more assurances that it will be a long term structure, that it won't suffer durability problems. We don't have strength problems. To us, high strength concrete is perhaps thirty, maybe even thirty-five megapascals at a year so strength is no issue with us but durability is. So, we need to work together, and one of the ways we can do that is certainly joint venture research, but within the industry.

Ole {Jensen} mentioned the fragmentation of the industry. We need to work together. I think I understand the profit situation and the industrial perspective there, but somehow we need to be collaborative rather than competitive. We need to get more people involved. Somebody mentioned Pavel {Trtik} being the youngest representative here. We need to work hard to get more of those kinds of folks involved in codes and standards. It doesn't seem to have the pizzazz that it used to have but, in fact, we need to work together. ASTM has a performance specification for cement. The Bureau of Reclamation is reluctant to do that because we don't have the performance tests to ensure the long-term viability. We need to work together.

We spent a lot of time, way back, when I think every person here talked about superplasticizers. In those codes and standards arenas, we spent a lot of time whether that was a high range water reducer or those were superplasticizers, a mere language issue. Currently, with its advent, albeit ten years ago, we're debating; is it self-compacting concrete or is it self-consolidating concrete? We spend time on those kind of issues and we need industrial help to push to the practical aspect of them.

Finally, we need to learn how to tell the world about our good things and get the public supporting us in the need for infrastructure research. There are efforts out there and, unfortunately, I've been involved in a couple of unsuccessful ones but going back several years to where ACI and Federal Highways tried to push a large research effort. We thought we had the administration's support because one of Clinton's big issues was infrastructure. Unfortunately, that went nowhere. Now, J.C. {Roumain} mentioned here the other day the Strategic Development Council, under the auspices of ACI, another opportunity for collaborative and joint venture researching. So we need to work together, undo the fragmentation that's there. Keep in mind that industry is there to make a profit, and that's okay, but somehow we need to work together to solve the issues that we the concrete industry face. Largely, this audience is an academia/industry area but government plays in that too, but in fact we have the same issues as when my administrators say; "Well, you've already solved the problems, don't worry about it anymore. We'll take care of it with a design and post-tension the ASR concrete."

Diamond: I'd like to change the tenor of the debate just a little bit and say something about tools. We've talked about modern scientific instrumentation, people have had lists of scanning electron microscopes and NMRs and all the rest of that stuff. That kind of equipment is available and is applicable for use in basic research in cements. We've also heard people like Juri {Gebauer} say that fundamentally our cement hasn't changed over the years. We've gotten very much more efficient at producing it but the benefits of basic research into the formulation of cement have never materialized. Someone pointed out very clearly one of the reasons this has not occurred is that to try out experimental formulations, we would need to perform a short-term production run and then it becomes cost prohibitive. This is not the way to go about it.

There used to be a small-scale kiln but I don't know if it still exists. I doubt it. PCA used to have one and this allowed the investigator to come in and do experimental runs. These things don't exist anymore. If we are to do research in the cement portion of the cement and concrete industry, we can't do it without the tools and those tools are not things we can borrow from the next lab down the corner. I wonder if indeed the cement industry is interested in this form of research, and why the need for facilities like this has been ignored for so many years.

Scrivener: Just some information to Sidney {Diamond} because this is a problem with which I was very heavily involved with during my time at Lafarge because I was in charge of the section making calcium aluminate cement. Calcium aluminate cement is a special cement so there we are actually confronted by projects where we were making new clinkers. We were probably the only research department doing that, mainly for refractory concretes and things like that. This is the eternal debate we had internally, and we did have the facilities and we still have a small pilot kiln. We know that doesn't produce something that's all that representative of the real production. It's horrendously expensive to run. I mean, to produce the eight tonnes of a new clinker we actually made I think from memory cost us, well it was a week's campaign and it cost us something like fifty thousand dollars, or twice that. I think you have to accept that's not the way forward in this day and age. I think we have to look much more at the fact that we do have most of the basic ingredients around and we can make materials much easily more by blending things together. I do think this is the way forward for new cementitious materials, rather than trying to play around with the clinker chemistry when we've got it pretty well sussed out and producing high volumes. I think what we need to think more about is taking the special materials we have available and putting them together. It's really not feasible to think about having a pilot plant in a laboratory. Even at Lafarge we couldn't make it work on a feasible basis.

Kirkpatrick: We have this manifesto that I think many of us would like to see fleshed out and we have a plan for doing that over the web but I guess I would like to refocus this discussion onto what I see are the two critical issues. The first one being clearly identifying the most important research issues, both on a kind of global scale, durability, and then underneath those, what the specific concerns are. So that's item one. Item two is how are we going to make use of this document and other parallel documents in a way that actually affects what it is we're trying to do.

I'm not sure I have an answer to all of that but at the very least what I'm willing to do is to see if I can identify those individuals in the organizations with which I have some contact, this would be say NSF and the Department of Energy, and other organizations like that. Also, senior academic people at the principal universities, that is department heads, deans, provosts, find out who those people are and to develop a way to get our ideas to them. So, I guess what I'm saying is I'm willing to see if I can at least identify some of those people and make some noise. And I would ask all the rest of you to think about who are the decision makers that we need to connect with and how do we best connect.

So, research issues. Many of you are much more qualified than I am to do that but let me go back to a point that I made yesterday. I really do think that, with some thought, we can develop a research agenda that is brought together by thinking of cement and concrete as a complex system in the way a physicist would talk about it. I think it's going to be a huge effort just to identify all the steps in that but I think what I'm willing to is to go back and put some though into that and try to understand how one does this mathematically and conceptually and see if we can come up with a way of putting some sort of global framework onto understanding concrete from the nanoscale through the structure scale.

Gebauer: I would like to come back to what Karen {Scrivener} and Sidney {Diamond} said about the difficulties in changing something in the Portland cement clinker itself. The main difficulty is the step from the laboratory tests or even pilot plant tests to industrial application, to industrial market application. This, as I showed in my presentation, is the most difficult and the most expensive step and is therefore very difficult to overcome and it is probably not possible to do so with Portland cement clinker at the present time. Probably the better way, as Karen {Scrivener} said, is to go first through blending and grinding improvements.

But, the best example of a failure of this approach is Pyrament, as previously mentioned. I mean, they didn't do the homework, and what is the homework? The homework is very expensive industrial production. It is not enough to make a pilot plant test. You have to work on the new type of binder for months and get experience. Portland cement has an advantage of a hundred years and if it is not done, this expensive exercise, then there is no chance for such new developments. I don't know how university research can help. If you want to make a change, how can the university help in this very difficult step? In the introduction of very innovative ideas. As I mentioned, I believe only in evolutionary development and this is slow, and realistic. Still, we have to try to do better, especially with the environmental constraints, to find maybe a binder which maybe will beat Portland cement one day. I don't think it will happen soon.

Mindess: I wonder, in the sort of limited time that we have, if we could return to the points that Jim {Kirkpatrick} made and if we can focus on that for a while. In particular, identifying the most pressing research needs that we see collectively and trying to think of ways in which we can somehow convince funding agencies and others to accept this view that we're going to come forward with. So I wonder if I can get some discussion around what's most important and get to the more fundamental question that I think I would like to have us address here.

Andrade: This point is precisely what we try in RILEM to stimulate at the national and regional levels. Everybody should start to discuss on the actors or on the agents, try to identify at the regional and national level and afterward try to put something together at the global level. In that case, we could have an influence afterwards, again, at the local level. I think that if we can think on this process over two years, we can meet in several places and try to finalize with a single work, but it should be large enough to accommodate many more local interests. I think that this is necessary. At the global level, the construction industry should demonstrate the benefits for society, and if society recognizes these it means that a certain percentage of the public finances should be in this area.

Concerning research priorities, there is one thing that I would like to say that I repeat many times to the steel industry, though when I address them they don't like to listen. I said that to the steel industry and they didn't like. The durability problems related to corrosion are not problems with the concrete. It is the steel. I mean, we really need the steel industry to solve half of the problems of concrete durability. Without them, without new kinds of steels and new surface treatments and new developments, we cannot evolve. For instance, in the very nice presentation by the Florida DOT, they presented a lot of effort but nothing about the steel because the only evolution has been epoxy coating. Since this failed, what about stainless steel, what about cathodic protection, what about many other possibilities because it is the steel that is the problem, not the concrete.

Bergin: I'd just like to respond a little bit to Carmen {Andrade}. The presentation was kind of short but we do have current research in stainless steel clad rebar. We currently use cathodic protection in many of our structures, especially in the Keys. I can, I would be happy to, give you the names the names of some people that you could contact if you are interested in some specifics on what we are doing. We also have a cathodic monitoring system that's tied into cellular phones so that the personnel at the corrosion laboratory can monitor these cathodic protections remotely. There is actually quite a bit of research in the reinforcing steel area that the department is doing. Unfortunately, I just didn't have time to include it in my presentation.

Thaulow: I would like to talk a little about how we in Denmark address this problem of getting public funding and getting cooperation between different companies. About ten years ago, we created a unity called the Danish Concrete Institute. It involved the three major contractors and the three major consulting engineering companies that went together to create it. There was a full-time person employed and there was a part-time managing director, which was me. The annual cost was about $100,000 to do this but because, once a month, the vice presidents of these companies could meet on neutral ground and discuss areas of common interests and define areas of research, we were able to get funding from different Danish agencies, including the Department of Transportation. They sponsored about three million dollars in research in high performance concrete but with a focus on the technology that the contractors needed in order to use it. I think this was really a great success. We were able to get European sponsorship because there was industry support, so in the final analysis it was maybe about ten percent of the total cost that companies put into it but because they joined forces we were able to get government funds.

Another function of the institute was to provide training courses and post-graduate education in all levels of civil engineering, including how to inspect structures. Also, we did courses in connection with big construction, like the Great Belt Link between Denmark and Sweden, where the contractor had to send his people to learn why there were specifications like there were or why it was important not to over-vibrate concrete. All of this was addressed. I think both the education part and the research part was a great success. Of course, being here, unfortunately, I'm not part of that anymore.

Scrivener: I just wanted to respond to what Jim {Kirkpatrick} was saying about the need to try and better clarify the physical chemical processes going on at all length scales. This may not be something that will be very easy to communicate to contractors and people like that but nevertheless I think it's a very necessary process to identify where we really have gaps because even things we think are fairly basic, there is a lot of very basic knowledge that we don't have. We may say first order problems have been solved, but it's quite clear that we have a lot of parameters to deal with. We see people using coefficients generated by Powers fifty years ago which were very good for what equipment he had available but the most simple evaluation shows that they're just not right for what we're doing today so I'd like to just support that claim and if he wants any help in that process, I'd be willing to help him out.

Forster: I think I'd like to get back to the question of prioritizing the research that we need to do in the cement and concrete area and I think I'd like to come back to is the group we have assembled here all-inclusive enough to really address it. I think it was pointed out, first of all, that we don't have any real contractors, constructors if you will, in the group and I think we certainly need their input. Those are the people that are going to have to work with whatever this material is we come up with. Secondly, we need the structural designers, whether it be actual buildings or bridges or pavements or whatever. What are their needs?

In pavements, for instance, high strength concrete is of no interest. As Jim {Pierce} mentioned, as far as dams are concerned or more massive structures, there's a maximum strength that he's interested in which isn't very high. The same thing applies to pavements, maybe it doubles the strength that he was talking about but it's still not what many of us consider to be high strength concrete these days. So, perhaps instead of the folks that are intimately involved with cement research and concrete research coming up with new advances, then at that point you decide, well what can we do with these new advances, you need a part of the input from the people that have to build with it and the people that design structures using this material. Well, what is the advantage or what are we looking for; thinner members, longer members, or fewer members that will take the same loads as the current structures that we're doing now?

Mitchell: I wanted to make a few points and it's a general problem that I think we have, instead of answering your question directly. We're in a period of drought now. I'm talking about the researchers. We saw a great harvest period a few years ago with the advent of high performance concrete. In every country, different research groups jumped on the bandwagon. That dried up, from the federal governments point of view of funding. So, I think we have to look at innovative ways of restarting things again.

I remember, I think it was Ole's {Jensen} curve of innovation, growth and stagnation. I mean, we could argue where we are on that curve but I think we would all be convinced that we're still in the growth period. The problem is that federal government funding has dropped off. That's a reality and the people in this room could change the picture. What we're trying to do in Canada is involve the construction companies. There's the Canadian Construction Association. As in all countries, construction companies are typically very small, ninety-five percent of them have less than twenty employees. It's the large ones that we're going after and, with the government backing us, a few universities are trying to convince the construction companies to put money into the federal funding agency in order to get matching funds to create new centers of research. I think that we ought to earmark the cement and concrete industries separately for the same thing. I think that if we did that we would get back on the bandwagon and we would all benefit from it, those in materials and structures. It was a wonderful time and I think we have to return to that and work out innovative ways of getting back to that.

Pierce: I would echo a thought put forward by Doug {Hooton} when he spoke, and perhaps considered somewhat of a heresy here in this crowd, but I think what we really need, at least from our perspective as an owner, is we need research into the system, not the components. We need the research to tell us how the concrete is going to perform. Yes, I can understand traditionally we have always looked at the components of cement, admixtures, aggregates but it is whatever that binder is or whatever the materials are and how it performs. We need to look at that kind of result moreso I believe than the component research.

I've been involved in both ASTM and ACI. In ASTM, for those of you who are familiar with that structure, there is committee C1 on cements and there is committee C9 on concrete and concrete aggregates. I've put forward the idea that we should even have committee C19, which would then combine C1 and C9. So far, that's not fallen on receptive ears but I think that is really the way we need to go. In terms of research priorities, there have been, like a lot of other papers that we have referred to, lots of concrete or cement or component research priorities identified but then we seem to fall flat. We get a long list of priorities but there's no follow through so we need to look at this in a sequential fashion. You may want to establish what is the highest priority. Again, I say from the government's perspective, that in fact we need to have durability so that the public is assured of its investment. We can't have people held up on the Sunshine Skyway bridge, we've got to have them always using the bridge and that sort of thing. The aspect of how we get that is really to me the big dilemma today.

Hooton: My experience with durability issues on construction sites is that it's the lowest person on the totem pole who has the biggest impact on durability. It's the truck drivers, it's the people placing the concrete, and it's the people dealing with the small details on construction projects that make or break the durability of the concrete. It's not the mix design. It's not the specifiers. It's the people on the project. One of the biggest hangups in our area against the use of high performance concrete for parking decks and bridge decks is the issue of plastic shrinkage cracking, which is easily solved with fog misting. You buy a power car sprayer and you have a guy standing upwind. As long as you explain to him why he's standing there upwind, and what the purpose of the project is. The only way these things succeed is if you educate the lowest person on the totem pole so that happens. When it doesn't happen, he's off looking the other way or having a smoke or whatever he's doing and the plastic shrinkage cracks appear. The project is destroyed with this expensive concrete. It doesn't matter what you put in it if the lowest person on the totem pole is not on board.

Jachimowicz: Maybe we can get back to the question of a manifesto. At the beginning, we had kind of a limited agenda. We didn't want to solve all the world's problems. We wanted to create a vehicle, to create a document that will raise the prominence of this area, that will allow us to talk to others in more exciting terms, that if I had a graduate student who is trying to choose which area to study, I can talk to him in words or concepts that will excite him. So that, if he had a choice to go into the biotechnology or to do research in cement based materials there would be a fifty-fifty chance, at least. We don't have a fifty-fifty chance right now. That's the challenge. That's how we conceived it. That's what we wanted to do, to create a template, a document, a blueprint that will guide this dialogue and that will communicate the importance, the excitement, the sophistication behind our work.

We are very frequently misunderstood. We keep using the term *low tech*, I don't know what low tech is. If you look at the technologies, the methodologies, the thoughts, the sophistication of thinking that goes into what we are doing, it is no less than in any other area. I think that is what we are lacking. We have to communicate that it is as difficult to create a new product, a new technology, a new concept in cement-based materials as it is to devise a new vaccine. I don't think that we will go very far by trying to convert this area to some sort of specialty materials. We won't convince anybody about this. I think that is not a very fruitful direction to go but what we can convince the public about, audiences about, is that this is a very sophisticated, a very complicated idea that needs the most input we can get. That was the concept behind trying to devise this document, not to solve everything. Another point that I want to make is that we don't think that it will take years to create it. We're not thinking in years, we're thinking in months.

Skalny: My main problem when we discussed this manifesto six or seven months ago for the first time was how we are going to sell the idea. Let's assume that we come up with a very nice scenario of what to do. How are we going to get it out to the industry, to the specifiers, to the structural engineers. You are right, Steve {Forster}, we do not have the proper people here and one of the problems is that I don't know them. I don't have enough contacts in that field that I could ask to say something. We tried to invite people we thought would be helpful, and they are, but obviously we missed some specialties. What we need in addition to technical help to make this list to specify what we need is help on how to sell it. I think the selling point is much more difficult than what we do here now. Like I learned during my career, research is easy compared to selling the research to management. This is a problem we have to address.

Zanni: Concerning how to attract students interested in cement and concrete, I think that environmental problems are very attractive to our students and not only in civil engineering departments, in chemistry and physics and everywhere. We can propose the use cement and concrete products for environmental things like, for instance, the embedding of waste. We have a lot of research on this point in France because we have nuclear waste, as do you. We have just finished with a European community program about the leaching of concrete and I think it's a very important thing, not only with nuclear waste, but with all industrial waste. Another point is the sealing of the waste, of petroleum waste, and in this case it must last for eternity. It's a question of durability for eternity. It's a big challenge and I think students might be interested by it.

Forster: I think she's hit on one of the keys. We've mentioned biotechnology. It's attractive, it's exciting to the students. How can we make cement and concrete research similarly attractive? The environmental issues are certainly one of the ways we can do that. Just saying you're going to work on cement and concrete and make it better, what's that mean? The young people coming up, they can't visualize how they're going to make a difference in the world doing that. You need to somehow relate that to the end product. What's going to happen in terms of buildings? Are they going to be totally different structures, pavements, whatever? The things that you're making out of the concrete, how are they going to be totally different and as a result change the world?

Thaulow: To follow up, I think what we need is a new buzzword. High performance concrete is no longer good for selling and I know that, at least in part of Europe, the new word is *green concrete*. This encompasses everything we have been talking about, using waste materials, reducing carbon dioxide emissions, but also life cycle cost. That includes durability, so suddenly durability would be in again under that heading of a green concrete because it's more environmentally friendly if it lasts longer. Maybe you can think of a better word, but I think that's where we're going and maybe this can attract young people because they're very conscious about the environment.

Ko: As, maybe not the youngest participant here but the younger generation, the buzzword for us was *sustainable environment*. Not only a green concrete but we are looking at even longer term goals to develop a sustainable environment. These can provide us with the approach from the materials side, the end product, with many concepts to convince the public that we are developing a technology to sustain our environment, to make the buildings more durable, to decrease pollution, to store our wastes in a better way to sustain future life, to sustain future generations. To continue the green concrete concept, maybe sustainable environment is really the end target.

Brandt: It seems to me that the answer, at least to that question, is relatively simple. To attract capable students to our field we just show them how complicated cement and concrete materials are and how relatively advanced are ways that should be applied to explain the processes, the phenomena and the results. This is very attractive for young people. From our experience, as soon as we showed that we have something to propose to them that is interesting, which is high technology, we got several quite clever young students to join us. This is probably simple, but how to address the founding bodies. It seems to me that we should address them through life cycle design, which includes durability, and through sustainability as an issue. Then we could persuade them that in building and repairing infrastructure we can propose a real input into civilization and then get enough funding for new research.

Trtik: I have to comment on Doug's {Hooton} remark that concrete is very much influenced by somebody who is not really well trained, that they have a really big influence on it, the people who place it and so on, people who are not well trained or lowly trained. In nanoseconds I had an idea, a comparison to my favorite sport, golf. What you get on your card in golf, it's not the result of your best shot, it's the result of your worst shot during the round of golf. It reminds me of the discussion between Sidney Diamond and Karen Scrivener, Jacques Lukasik and myself. Sidney Diamond would like to improve the first shots, work on them. Karen Scrivener, Jacques Lukasik and I, we are the guys who would like to investigate better drivers, better clubs, better balls. Both ways probably lead to a better result. The question is; where should we put more energy, into improving our worst shots or getting the better equipment for it. I really don't know. It's the question for discussion. The second point which I would like to make is, I can't agree more with Leslie {Ko} that environmental issues are the things which, I think for my generation, are really important and I said it already yesterday.

Gebauer: Referring to the research manifesto, I expect that from this meeting we can really focus and set some directions for our future research activities. Even if we do not finish here, we are missing some of the people who are involved in our business, but from the materials science side certainly the leading people are here and we should at least try to formulate the important topics. We heard many of them here and we need to make some contribution to this manifesto, to make set some direction for research in order to really make our research world more attractive to the students, the universities and also to industry. This, I think we should be able to contribute. We heard some topics, sustainable development or microstructural design, and I don't know if we still have the time for this but some time later we should really try to focus and make some plan of action plan to do something and not just talk.

Mindess: I wonder, if perhaps I should ask people like Juri {Gebauer} and Jacques {Lukasik} a question. Your companies spend quite a lot of money on advertising, one way or other, with products, brochures and all of that stuff. Have you ever asked your advertising agencies or your advertising people to try and actually structure a campaign about raising the profile generally with government or with funding agencies instead of doing this essentially to compete amongst yourselves. I mean, like anything else, advertising requires expertise and none of us are particularly good at it. That's not our training. I'm just wondering if there is some professional help that you already have that you might turn toward in other directions a little bit.

Gebauer: It's just that we are in a business where we are not doing a lot of advertising compared to other products. We are advertising to our customers, only direct customers, but it certainly may be a good idea to go in that direction and try to make this advertisement more public. There are some activities in Germany, for instance, where they are advertising concrete as a material on the TV, at various shows and distributing pamphlets but it's not done enough I'm sure.

Lukasik: We do it on a rather individual basis, as a company not as an industry. We have some channels for general promoting of our products through professional associations, nationally and Europe wide. So, maybe we should do it more specifically.

I would like to comment on Pavel's analogy to golf. I would like to return to this best shot idea and make another analogy. Let's go back to the automobile, the auto industry again. There's this Formula One racing and I think this is the type of offensive approach to research in which technology is pushing one way and perhaps so far ahead of everything else that there are spinoffs that go back to help those worst shots at the low level. From F1 cars there are many spinoffs in our everyday cars and when we work on products like Ductal, we learn plenty of things that we can use in those thirty megapascal concretes about placement, self-compacting concrete and so on. So I encourage everybody who can invest in this F1 type of research to do a little bit of it. It pays off in the long term, but it needs a lot of persuasion of your top management to spend resources and allocate some money.

Drakos: I want to go back to a couple of comparisons that were made earlier. The first one was comparing concrete to microchip technology. For one, you're presenting concrete as a commodity but think about who you are selling it to. I hear the term buzzword, what is the buzzword? Somebody said green concrete. Well, take this opportunity of being in Florida, a place that is often hit by hurricanes but still people are using lumber to build houses. You have a hurricane every three or four years that would, if it passed through this lovely island I'm sorry to say, destroy everything. So, what is the opportunity for the industry here. Come up with a concrete product which will be durable, meaning withstand a hurricane, recyclable maybe to give it the green aspect, and then cheaper or as cheap as lumber houses. There is a market that where can go to the end user. When you come up with a new product that applies only to contractors, you're not planning for the end user, which is me. I don't get to deal with it. It's not like a microprocessor which I use every day to do my job and has a half life of two or three years and is then obsolete. You're talking about a concrete building that I'm going to design for fifty years so there's no real comparison between the two. So, if you want to do the cool thing or the thing that people are going to start turning some eyes on, maybe target the end user a little bit, the homeowner. Do something that would generate some concern about your product.

Skalny: I wanted to respond to what Juri {Gebauer} and some others said. I think that this idea of a manifesto was to come up with some good ideas today, but this is not the end of it. Now, Felek {Jachimowicz} may tell you more about it. We are already working on a website which will be available to all of us plus other people worldwide to come up with these ideas which Jim {Kirkpatrick} hopefully is collecting. I don't want only Jim and those of us who had this idea to be involved, but think that everybody who is willing to be active and respond in time or suggest other people from structural engineering down to nanotechnology. Everybody should feel free to send us material, not too long so that we can organize it into something useful. Felek {Jachimowicz}, is that what you visualize on how we follow up, because it's obvious that we are not going to finish today? This is not the end of it. This is the beginning.

Boyd: I think Jan {Skalny} had a good point when he spoke of the need for advertising. This aspect is really lacking in our field. I've heard a number of comments about trying to attract new students or new people into the field or generating interest at the government level for increased funding. Unfortunately, there is this negative aspect that keeps coming up about it being a low-tech industry, which prevents such people from being interested. That's simply not true. Look at the testing techniques we commonly use, SEM, XRD, MRI. We're talking about very high end testing and research but nobody knows that outside the actual research field. We need the advertising to get this information out to the general public, to prospective students, and funding agencies.

Skalny: I would just like to just remind everyone that in my view the reason that materials science departments do not work on concrete and cement, but rather on steel and ceramics, is because that's much easier. Concrete is much more complicated than working on ceramics. Those are crystalline materials with nice structure and we handle them differently. Concrete is a complex material, fifty chemicals in a clinker, each of them reacts in a different way, many different mechanisms. It's very complex. That's the reason that they are not going to do it.

Kirkpatrick: I guess I have two questions. One is on the advertising business. In the United States, a company called BASF® advertises very heavily on television and their every ad has as its theme "We don't make a lot of the products you use, we make a lot of the products you use better." That's their thing. I don't understand what they're getting out of this but they've been doing it for decades, a decade anyway that I can think of. I guess one of the questions that I would put to the proper folks here is to try and find out what that kind of advertising to the public, to people who don't actually consume the product that the company makes, what's the purpose behind it? Second question I have for all of you. In any of your countries, any of your states or wherever you are, are there any situations where the contractors and the individual workers who actually place concrete have to go through any sort of licensing program to ensure that they know enough to be able to do the job right.

Lukasik: Just on the second point Jim {Kirkpatrick}. For Ductal yes. In South Korea, on the particular bridge that I showed, there was special preparation, special training. I mean, we didn't let just anybody handle the precasting operation before we figured out whether they were trained enough and they wouldn't do, as we say in French "n'a pas de quoi".

Boyd: We have trained welders, we have trained carpenters, we have trained people for virtually every other building material except concrete it seems, where a contractor is going to pick up a laborer off the nearest street corner to do his work for him. It seems rather ironic to me that the most complex of our building materials actually requires the least amount of training in today's industry.

Mindess: Actually, that used to be one of my functions, to drive to what back then was called the unemployment insurance agency where there would be just a whole bunch of people milling around. The question was not "do you know anything?" but "do you have a shovel?"

Olek: I just wanted to make one comment related to, shall we say, casting the net to attract supporters for what we want to do. As I observed the activities in the ACBM center, it was very apparent that the only supporters of activities aimed at basic research are the cement companies, despite the fact that I know there were individual contributors from the precast industry, for example, to support some of the research programs. We weren't successful in getting them involved in supporting research on a more global scale.

There are companies here, Lafarge being one of them, that have their own precast or ready mix concrete operations. In addition to cement companies, we do have a large market segment involving ready mix concrete and precast producers that are typically not represented in the area that we are dealing with. My suggestion would be that, as we are preparing the manifesto and as we are trying to get a lot of people involved, we shouldn't overlook those components. From talking to some of them, I think that they always feel sort of left out of the mainstream and might be willing to step up and help us create a campaign that will carry the message of the importance of that topic.

Gebauer: About ready mix producers, they are simply not organized as well as the cement industry and they are not able to, I think, contribute to these research activities. Actually, they are considered more as a transporting company dealing with logistics and economical transporting. But, the main problem is that they are fragmented and not as organized as the cement industry. The precast industry is different. They are sometimes even involved in high tech situations and are able to contribute to the needs of our customers. I just wanted to point out that ready mix is a very special business and on average almost fifty percent of cement in the world goes into ready mix.

Mindess: We're drawing quickly to a close here. I wonder, first of all, whether the people in this room are, in general, supportive of going ahead with the manifesto in one form or another. Is this something that people are willing to devote some time and intellectual energy to? I don't expect you to sort of leap up and say "I declare" but I'd like to think that within this room there is some interest in that, though I see Doug {Hooton} is shaking his head. I think the hope is that we can use this document and others such as the 2030 vision that J.C. {Roumain} kindly brought to us to try and influence funding agencies, and perhaps government agencies and industry, to try and focus a little bit more on what kind of research is funded. I mean, there's a lot of money spent on research by organizations like NSF. I'm not sure that it's directed in the most sensible way. I'm sure it's not, in fact. I think that documents such as this can help to refocus, to some degree at least, the kinds of things that we do. Perhaps I could give Felek {Jachimowicz} the last word since in many ways he is the father of this document and perhaps he could say something in closing about what he hopes to do with it.

Jachimowicz: What are we going to do with it? To use it to communicate. To put more excitement into what we do. For example, somebody is applying for a grant, he or she will be using the language, the formulae, the concepts from the manifesto. Look, go to NSF, got to the granting agency. This is a document that was composed, put together by the experts in the industry, by the experts in the field. My research, my grant, my proposal is aligned very well with it. This is just one of the examples. If we come up with a good document, we will try to talk about it in many different forums, to many different audiences, as much as possible.

Hooton: I guess one of my problems with this is that the needs change. There are some big long-term needs but in general the needs change. You're saying, in this example you just gave, yeah okay the research I proposed is on the approved list of the manifesto so it has a better chance of being funded. I've seen that used to prevent funding because somebody comes up with a new idea that's not on the list and they say, I'm sorry but it's not on the list, it's not on the radar screen. I see that as a double edged sword and I guess, as someone who I think keeps his head to the ground and knows what's going on, I don't necessarily see the advantage of that unless we only use it to attract the attention of the politicians to put funding back in the area.

Skalny: I just wanted to again remind everybody that this is a living document. We are not going to put out a piece of paper and then fall asleep. We hope that this will continue in different forms, at different meetings. We have to talk to the Strategic Development Council. RILEM has a committee that should cooperate and maybe jointly with everybody else we can push the technology at least this amount ahead. If we achieve that, we have already contributed. I don't think this will be perfect. I'm sure it is not but it's worth the try to add a little more pressure to what we need in that direction. So I don't expect complete success but small progress is also progress.

Mindess: I think with that, it's probably time to draw this meeting to a close. I think we've had a lot of fun and some very good ideas over the last few days. I hope that when we return here next year, at this time [lots of applause and shouts of support], we will actually be able to report some progress. I think that although many of us might be cynical about what we can achieve, the danger in not doing anything at all is far greater than the danger of trying and not succeeding so I would urge you all to participate, if you wish, in an effort to try and bring cement and concrete research to the forefront again. Thank you all for coming and we'll see you next year.

Photo 47: Anna Maria Workshop 2001 Participants

KEYWORD AND AUTHOR INDEX

Printed and bound by CPI Group (UK) Ltd, Croydon, CR0 4YY

15/09/2026

14962409-0004